Conceptual Development Activities

Bothell, WA • Chicago, IL • Columbus, OH • New York, NY

Author

Sharon Griffin
Professor of Education and Adjunct Associate Professor of Psychology
Clark University
Worcester, Massachusetts

Building Blocks Authors

Douglas H. Clements
Professor of Early Childhood and Mathematics Education
University at Buffalo, State University of New York, New York

Julie Sarama
Associate Professor of Mathematics Education
University at Buffalo, State University of New York, New York

Reviewers

Lynn G. Berger, M.Ed., *Instructional Support Teacher, Exceptional Student Services,* Orange County Public Schools; **Mary Craig,** *Program Specialist, ESE Department,* Lake County Schools; **Gail Filson,** *Classroom Support Specialist for Title I,* Volusia County Schools; **Sharon Lock,** *Instructional Support Teacher, Exceptional Student Services,* Orange County Public Schools; **Beth Phillips,** *Intellectual Disabilities Coordinator,* Polk County Public Schools; **Sam Thompson,** *Community Based Instructor,* Leon County School District; **Virginia Weidner, Ed.D.,** *Program Specialist ESE—Curriculum,* Leon County Schools

SRAonline.com

Send all inquiries to:
McGraw-Hill Education
4400 Easton Commons
Columbus, OH 43219

ISBN: 978-0-02-113803-6
MHID: 0-02-113803-6

Printed in the United States of America.

2 3 4 5 6 7 8 9 QDB 17 16 15 14 13 12 11

Acknowledgments

Development of the ***Number Worlds*** program was made possible by generous grants from the James S. McDonnell Foundation. The author gratefully acknowledges this support as well as the contributions of all the teachers and children who used the program in various stages of development and who helped shape its current form.

Building Blocks was supported in part by the National Science Foundation under Grant No. ESI-9730804, "Building Blocks—Foundations for Mathematical Thinking, Pre-Kindergarten to Grade 2: Research-based Materials Development" to Douglas H. Clements and Julie Sarama. The curriculum was also based partly upon work supported in part by the Institute of Educational Sciences (U.S. Dept. of Education, under the Interagency Education Research Initiative, or IERI, a collaboration of the IES, NSF, and NICHHD) under Grant No. R305K05157, "Scaling Trajectories and Technologies" and by the IERI through a National Science Foundation NSF Grant No. REC-0228440, "Scaling Up the Implementation of a Pre-Kindergarten Mathematics Curricula: Teaching for Understanding with Trajectories and Technologies." Any opinions, findings, and conclusions or recommendations expressed in this material are those of the authors and do not necessarily reflect the views of the funding agencies.

Contents

Response to Intervention

RtI represents a profound change in the way educators help struggling students. It provides a different approach to teaching math to at-risk children, and requires a coordinated effort between administration and staff to be successful. With its research-based curriculum and extensive field testing, ***Number Worlds*** supports RtI and helps schools meet their academic objectives. Just as RtI encourages work with at-risk students early on, ***Number Worlds*** is the only math intervention program with a prevention program for Kindergarten and Grade 1 students. ***Number Worlds*** gives at-risk students the confidence and skills to excel in math.

Building Blocks, featured in ***Number Worlds*** Levels A–H, provides engaging research-based computer activities and a management system that guides children through mathematical learning trajectories.

Building Blocks software that supports ***Number Worlds*** concept development was built on research conducted in a well-defined, rigorous, and complete fashion.

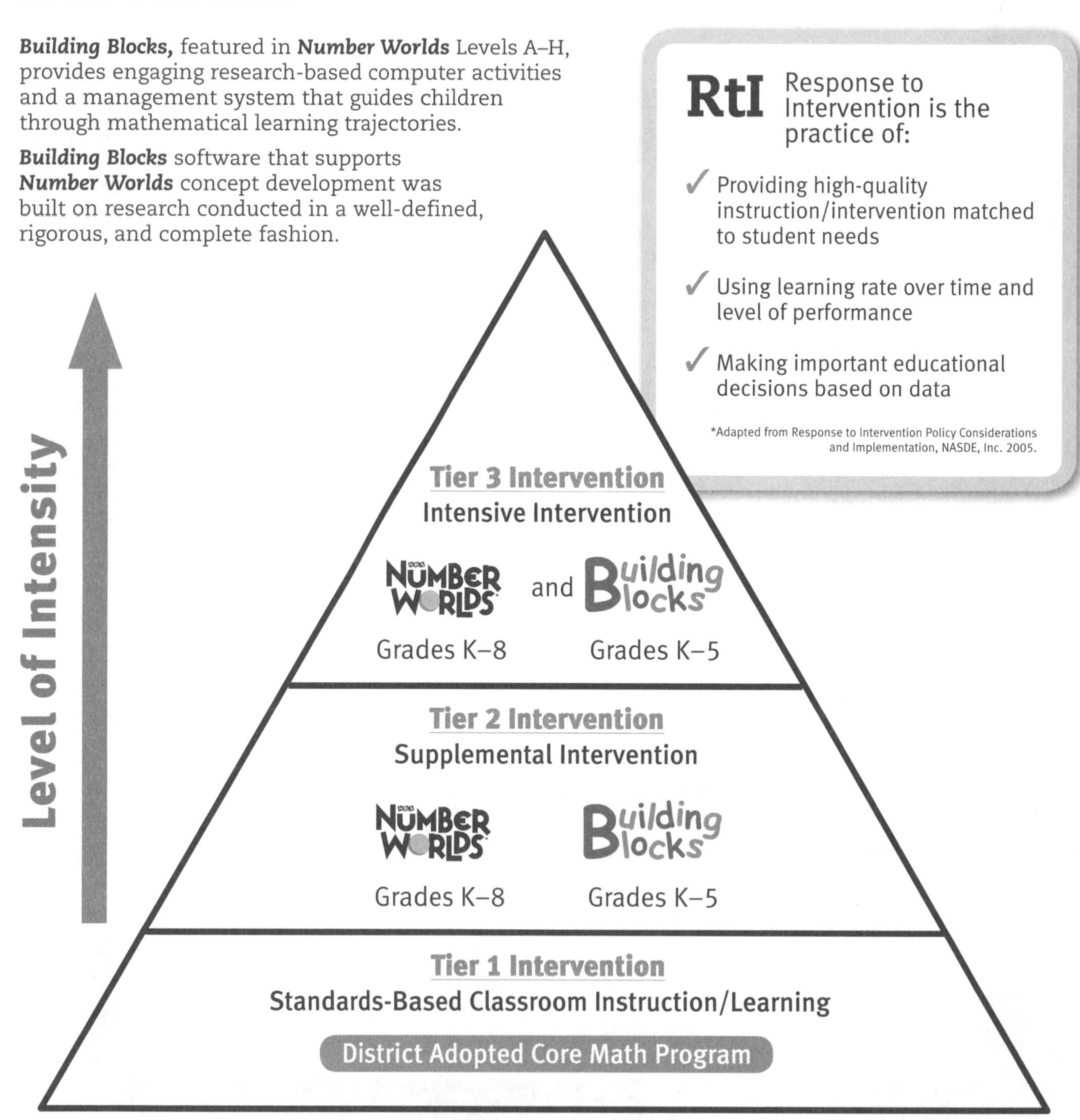

Number Worlds and Exceptional Students

Number Worlds provides a wealth of explicit and embedded strategies to provide rich instruction for exceptional students. They ensure access to the general curriculum at each grade level. Conceptual Development Activities describe the knowledge and skills required at each grade level, and are written within three levels of complexity: Independent, Developing, and Emerging, with the Emerging level being least complex.

Conceptual Development Activities Model

The repertoire of research-based instructional strategies and resources provided in ***Number Worlds*** ensures success for exceptional students.

Number Worlds strategies...

- Guided discussion and reflection supported, encouraged and developed in each lesson
- Skill building, strategy building, and hands-on activities in every lesson
- Extended skills practice provided in student workbook exercises, practice blackline masters, ***Building Blocks*** software, and ***eMathTools*** software
- Academic vocabulary and Access vocabulary identified for each unit and lesson
- Involvement of parents encouraged by Letters to Home for every unit
- Multiple ways to check for student understanding, including Warm-Up exercises, Informal Assessments, Rubrics, Progress Monitoring, and Student Workbook exercises
- Demonstration of understanding observed in discussion, hands-on activities, response activities, and game play
- Opportunities for students to communicate their understandings orally and in writing using Guided Discussion, Reflect, and student workbook exercises

Number Worlds Research

Author Research

Number Worlds authors have made significant contributions to mathematics education research. This research forms the foundation of the program.

> "Research-based, scientifically validated interventions/ instruction provide our best shot at implementing strategies that will be effective for a large majority of students."
>
> *Response to Intervention Policy Considerations and Implementation, National Association of State Directors of Special Educations, Inc. ©2006, p. 20*

__Number Worlds__ was developed to expose and develop children's understanding of the three worlds of mathematics: quantity, number, and symbols. Children develop understanding by exploring five different ways numbers and quantities are represented.

In ***Object Land*** students explore the world of counting numbers by counting and comparing sets of objects or pictures of objects. In Object Land you might ask:

- **How many or few do you have?**
- **Which is bigger or smaller?**

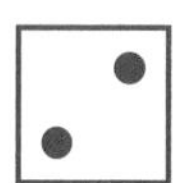

In ***Picture Land*** numbers are represented as sets of stylized, semi-abstract dot-set patterns, such as in a die and also as tally marks and numerals. In Picture Land you might ask:

- **What did you roll/ pick?**
- **Which has more or less?**

In ***Line Land*** number is represented as a position on a path or a line. The language used for numbers in Line Land refers to a particular place on a line and also to the moves along a line. In Line Land you might ask:

- **How far did you go?**
- **Do you go forward or backward?**

In ***Sky Land*** number is represented as a position on a vertical scale, such as on a thermometer or a bar graph. In Sky Land you might ask:

- **How high or low are you now?**
- **Who is above or below?**

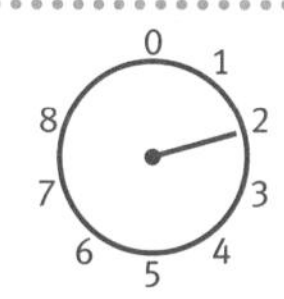

In ***Circle Land*** number is represented as a point on a dial, such as a clock face or a sundial. In Circle Land you might ask:

- **How many times did you go around?**
- **What did you land on?**

Number Worlds has been developed and refined since the mid-1980s and has shown proven results through years of rigorous field testing. These results show that students who performed below their peers surpassed the performance of students who began on-level with their peers simply by participating in the **Number Worlds** program.

As the figure shows, the magnet school group began kindergarten with substantially higher scores on the Number Knowledge Test than those of children in the Number Worlds and control groups. The gap indicated a developmental lag that exceeded one year, and for many children in the Number Worlds group, it was closer to two years. By the end of the kindergarten year, however, the Number Worlds children had narrowed this gap to a small fraction of its initial size. By the end of second grade, the Number Worlds children actually outperformed the magnet school group. In contrast, the initial gap between the control group and the magnet school group did not narrow over time. The control group children did make steady progress over the 3 years; however, they were never able to catch up.

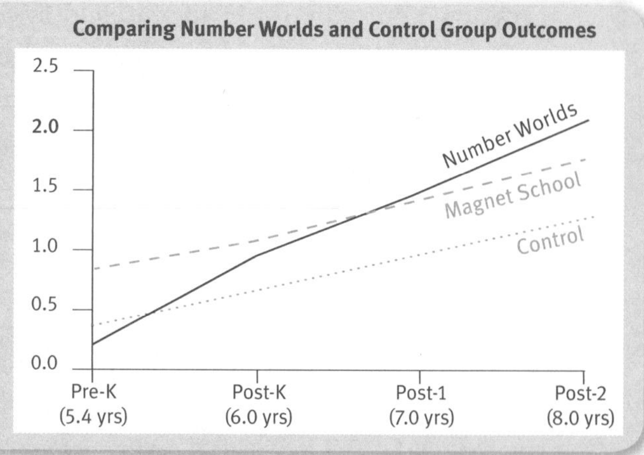

Griffin, Sharon. (2005) Fostering the Development of Whole-Number Sense. How Students Learn: Mathematics in the Classroom. Washington, D.C.: The National Academies Press.

Conceptual Development Activities

Activity 1

Independent Objective

Students use objects and number names to represent quantities to 5.

Display five counters or objects in a group. Pull one object toward you, ring the object with your finger, and say "one." Repeat with *two* through *five*, each time saying the number's name and ringing that number of objects. After each set, pull all the objects together before identifying a new set.
Give each student five counters or objects. Continue by asking students to create and identify numbers of objects, first in numerical order and then in random order. Remember to have them pull all the objects together before forming another set.

- **Make a set of two.**
- **Did you make a set of one or a set of two?**
- **Show me/Tell me how many.**

Repeat with quantities from 1 to 5.

Lesson Follow-Up
If students have difficulty completing this activity, have them complete Activity 4.

Activity 2

Independent Objective

Students develop one-to-one correspondence and compare sets to 5.

Show two red counters in one hand and two blue counters in the other hand, and ask,

- **Do I have the same amount of counters in each hand?**
- **How do you know?**

If students suggest matching the sets, ask a volunteer to do it. If not, show students how to match the sets one-to-one. Then ask,

- **Does each set have the same amount?**

Show two more sets, one red set with several more than the other blue set, and ask,

- **Do the two sets have the same amount?**
- **How do you know?**

Again, use one-to-one correspondence to check the answer. Ask questions such as

- **Show me/Tell me which set has more.** (Point to the larger set). **This set has more than** (point to the smaller set) **this set.**
- **Show me/Tell me which set has less.** (Point to the smaller set). **This set has less than** (point to the larger set) **this set.**

Repeat the activity with quantities from 1 to 5. Then ask volunteers to make the sets and ask questions.

Lesson Follow-Up
If students have difficulty completing this activity, have them complete Activity 5.

Conceptual Development Activities

Activity 3

Independent **Objective**

Students solve problems involving simple joining and separating to 5.

Display 2 blocks or counters.

- **How many?**

Put one more with the set, and tell students what you are doing.

- **How many did I add to the set?**
- **How many in all?**

Give each student five blocks or counters. Tell them to make a set of one counter.

- **How many?**

Then have them put another counter with it, and say what they are doing.

- **How many did you put into the set?**
- **How many in all?**

Repeat with a variety of sets, using quantities from 1 to 5.

Next, display five blocks or counters.

- **How many?**

Take one away.

- **How many did I take away from the set?**
- **Now how many are there in all?**

Have students form a set with a specific number of counters. Then have them take away a specific number, and follow the same line of questioning.

- **How many?**
- **How many did you take away from the set?**
- **Now how many are there in all?**

Lesson Follow-Up

If students have difficulty completing this activity, have them complete Activity 6.

Activity 4

Developing **Objective**

Students use objects and number names to represent quantities to 3.

Give each student three counters, and model counting as you place each counter in front of a student. Ring the counter with your finger, and say "one." Continue with *two* and *three*, ringing the appropriate amount each time. After you've distributed the counters, ask the students to use their counters and to repeat what you do and say. Display three counters. Pull one counter toward you, ring the counter with your finger, and say "one." Repeat with *two* and *three*, each time ringing the counters and saying the number's name. After each set, pull all the objects together before identifying a new set. Continue by asking individuals to make sets and name them, first in numerical order and then in random order. Provide help as students need it.

- **Make a set of one.**
- **Did you make a set of one or a set of two?**
- **Show me/Tell me how many.**

Repeat with quantities from 1 to 3.

Lesson Follow-Up

If students have difficulty completing this activity, have them complete Activities 7 and 8.

Conceptual Development Activities

Activity 5

Developing **Objective**

Students use one-to-one correspondence to count sets to 3.

Display three stuffed animals, puppets, or boxes. Have available three or four small toys or counters. Tell the students that each animal needs a toy, and ask,

- **How can we make sure that each animal has a toy?**

If students suggest something such as "Give each animal a toy," ask a volunteer to do it. If not, show students how to count the objects by matching the sets one-to-one. Then ask,

- **Does each animal have one toy?**

Take the toys from the stuffed animals. Then put two toys near the animals, and ask,

- **Will each animal have a toy? How do you know?**

Ask a student to match the toys to the animals. Provide help as needed. Then ask,

- **Does each animal have one toy? Are there more toys or more animals? Are there fewer toys or fewer animals?**

Repeat the activity with one toy and three animals.

Lesson Follow-Up

If students have difficulty completing this activity, have them complete Activity 9.

Activity 6

Developing **Objective**

Students solve problems involving simple joining to 3.

Display one block or counter.

- **How many?**

Then add one more to the set, and explain what you are doing.

- **How many did I put into the set?**
- **How many in all?**

Give each student three blocks or counters. Tell them to make a set of one counter.

- **How many?**

Then have them put another counter with it and tell you what they are doing.

- **How many did you put into the set?**
- **How many in all?**

Repeat several times with different combinations of counters. After students are able to join the sets, work with taking away from sets. Display three blocks or counters.

- **How many?**

Take one away.

- **How many did I take away from the set?**
- **Now how many are there in all?**

Have students form a set with a specific number of counters. Then have them take away a specific number, and follow the same line of questioning.

Lesson Follow-Up

If students have difficulty completing this activity, have them complete Activity 9.

Conceptual Development Activities

Activity 7

Emerging **Objective**

Students indicate desire for more of an action or object.

Display several small classroom objects, such as counters, blocks, or cars. Set one in front of yourself, and say,

- **Here is a counter. I want more counters.**

Take two or three, and put them with the other counter. Say,

- **I took two more counters.**

Then ask the student to take or point to a counter. Continue the conversation with questions such as

- **What did you take/What did you point to?**
- **Show/Tell me where you see more counters.**
- **Do you want more counters?**

Repeat this activity on several occasions until the student is comfortable with the routine.

Lesson Follow-Up

If students have difficulty completing this activity, do not use number words. Work only with *more* on several occasions, using classroom objects (more books, more pencils, or more chairs). As often as appropriate, reinforce the concept of *more* with all students.

Activity 8

Emerging **Objective**

Students indicate desire for no more of an action or object.

Display several small classroom objects, such as counters, blocks, or cars. Set one in front of yourself, and say,

- **Here is a counter. I want more counters.**

Take two or three, and put them with the other counter. Say,

- **I will stop. I don't want any more counters.**

Then ask a student to take or point to one counter. Continue the conversation, eliciting *more, no, yes, stop,* depending on the student's responses.

- **Show/Tell where you see more counters.**
- **Do you want more counters?**
- **Shall I stop?**

Repeat this activity on several occasions until the student is comfortable with the routine.

Lesson Follow-Up

If students have difficulty completing this activity, focus on one concept per activity. For example, work only with *more* or *stop* on several occasions. As often as appropriate, reinforce the concepts of *more* and *stop* with all students.

Conceptual Development Activities

Activity 9

Emerging Objective

Students use language such as *enough, too much,* or *more* to solve problems involving small quantities.

Use counters or familiar classroom objects with this activity. Show a set of one counter and another of two counters, and say,

- **Here is one counter. I need two counters. You need two counters. Which group has enough?**

To correct, show your counter again, and say,

- **Here is one counter. I need two counters. Two is one more than one. I do not have enough. I need one more. Show me one more counter.**

Repeat with one, adding two. Give the student a counter, and say,

- **Here is one counter. Show/Tell me how many counters you have.**
- **You need two counters. Do you have enough? Do you need more?**

Next, show two counters as you say,

- **Here are two counters. I need only one. I have too much. I will take away one counter.**

Repeat the activity with various small quantities, and have students show/tell whether they have too much, enough, or need more.

Lesson Follow-Up

If students have difficulty completing this activity, continue working on non-verbally comparing sets of one, two, and three, using one-to-one correspondence. Help the students identify which set has more or fewer.

Activity 10

Independent Objective

Students sort objects by single attributes such as shape and size.

Give each student a circular object. Have students trace around the shape with a finger. Then ask them to tell you about the shape. Give each student several more circular objects of various sizes to examine and tell how they are alike and different. Have the students put the circular shapes to the side.

Follow the same procedure with triangular shapes. Then ask the students to put the two sets together and mix them.

Hold up a circular shape, and say,

- **Make a set of circular shapes.**

If students have difficulty, point to a circular shape, and discuss the characteristics. Have the student find shapes "like this one" and put them next to each other.

Repeat the procedure with triangular shapes.

Finally, have students put two shapes side by side or on top of each other and tell which is bigger and which is smaller.

Lesson Follow-Up

If students have difficulty completing this activity, have them complete Activity 15.

Conceptual Development Activities

Activity 11

Independent Objective

Students match and name two-dimensional shapes such as circle and square.

Give the students a square, a triangular, and a circular attribute block or cutout. Ask:

- **How are they alike?**
- **How are they different?**

If necessary, have the students trace the edges of the cutouts and identify sides and corners. They should notice that one block does not have corners.

Have students take turns holding up an attribute block so the other students can name its shape. Display several of each type of attribute block. Then have a student find another attribute block that is the same shape.

Lesson Follow-Up

If students have difficulty completing this activity, have them complete Activity 16.

Activity 12

Independent Objective

Students match three-dimensional objects such as balls (spheres) and blocks (cubes).

Have two or three sizes of spheres and cubes for demonstration. Give each student several balls and blocks of a variety of sizes and colors. Tell the students,

- **I am going to hold up an object. You look at your objects, and make a set of those that are shaped like mine.**

Hold up a sphere, and say,

- **Make a set of all your objects like this one.**

Repeat this procedure several times.

Next, hold up one of the blocks, and say,

- **Here is a block. Show me/Tell me where you see something else in the classroom that is shaped like a block.**

Repeat this procedure several times with various shapes in a variety of sizes and colors.

Lesson Follow-Up

If students have difficulty completing this activity, have them complete Activity 17.

Conceptual Development Activities

Activity 13

Independent **Objective**

Students identify shapes in the environment such as circle and square.

Make sure there are several objects with square and circular shapes in the room. This could include floor tiles, clock faces, windows or windowpanes, can lids, stool seats, postage stamps, and notepads. Also have circular- and square-shaped attribute blocks or cutouts for the students.

Give each student a square attribute block. Have the students trace the edges of the attribute block and identify sides and corners. Ask them to tell what they know about square shapes. Then give each student a circular attribute block, and repeat the procedure above.

Next, divide the group into two teams. Have one team look for things shaped like a circle and the other for things shaped like a square. After a few minutes, ask the students to point out and name the objects they found. If necessary, clarify their understanding.

Lesson Follow-Up

If students have difficulty completing this activity, have them complete Activity 18.

Activity 14

Independent **Objective**

Students use words such as *in, out, up, down, top, bottom, on,* and *off* to identify spatial relationships.

Have available a box with a lid and an object that will fit in the box. Use these to demonstrate the position words *in, out, up, down, top, bottom, on,* and *off*. Keep in mind that a position word tells where a thing is in relation to another thing.

First, give a series of commands to the group. Keep the commands simple so you will know when to stop and demonstrate. For example:

- **Put your hand on your knee.**
- **Take your hand off your knee, and put it on top of your head.**
- **Stick out your thumb. Pull in your thumb.**
- **Put your book on the table. Take your book off the table.**

Next, give instructions to individuals.

- **Put your hand on the top of the box.**
- **Put your hand on the bottom of the box.**
- **Put the pencil in the box.**
- **Take the pencil out of the box.**
- **Hold the pencil down.**
- **Hold the pencil up.**

Lesson Follow-Up

If students have difficulty completing this activity, have them complete Activity 19.

Conceptual Development Activities

Activity 15

Developing **Objective**

Students sort common objects by size.

Display several strips of card stock of two sizes that look alike except for size. Ask a student to put the strips into two groups. If the student sorts the strips by size, say,

- **Tell me about your groups.**
- **Which group has long pieces?**
- **Which group has short pieces?**
- **How do you know?**

The student should place the groups beside each other to show that one group has strips that are longer.

If the student did not sort by size, say,

- **Tell me about your groups.**
- **Let's try to make two new groups. Put all the pieces together.**

Place a long strip next to a short strip. Say,

- **This piece is long. Point to it. The other is short. Point to it.**
- **Some of the pieces are long, and some are short. Make a group with long pieces and a group with short pieces.**

Repeat the activity with other classroom objects of different sizes.

Lesson Follow-Up

If students have difficulty completing this activity, have them complete Activity 20.

Activity 16

Developing **Objective**

Students identify square objects or pictures when given the name.

Before the lesson, make sure there are several objects with a square shape scattered throughout the room. Give students a square pattern block or cutout. Have the students trace the edges of the block and identify sides and corners. If necessary, run your hands along the sides and corners, saying "This is a side. This is a corner." When you have traced all four sides and corners, ask,

- **How many sides does this block have?**
- **How many corners?**
- **This block has a square shape. All square shapes have four sides and four corners, and all the sides are the same size. All the corners are the same size.**
- **What does every square shape have?**

Go on a "square-shape hunt" with the students. Stop occasionally, and ask students to point to an object with a square shape. Say to the students:

- **How many sides does it have?**
- **Are the sides the same size?**
- **How many corners does it have?**
- **Are the corners the same size?**
- **What do we call that object?**

Lesson Follow-Up

If students have difficulty completing this activity, have them complete Activity 20.

Conceptual Development Activities

Activity 17

Developing **Objective**

Students identify three-dimensional objects such as a block (cube) or ball (sphere).

Display several common objects that are shaped like a cube or a sphere, such as balls, oranges, round dried peas or beans, stones, marbles, boxes, blocks, number cubes, or a cube-shaped tissue box.

Give an object to a student, and say,

- **Look at this carefully. What is it?**

If necessary, name the item in a complete sentence, ask the student to say the name, and ask the group to say the name. Repeat the process until all the students have had a chance to name an object. Then point to an object, and ask the group to name it. If necessary, name the item again in a complete sentence, and have students point to the item named.

Hold up a cube, and say,

- **Here is a cube. Show me/Tell me where you see something else shaped like this.**

Give each student the opportunity to name the object and tell about it. Then repeat the activity with other objects.

Lesson Follow-Up

If students have difficulty completing this activity, have them complete Activity 21.

Activity 18

Developing **Objective**

Students identify square shapes in the environment when given the name.

Identify several objects with square and circular shapes in the room, such as floor tiles, clock faces, windowpanes, can lids, pattern tiles, postage stamps, and notepads. Also, have circular- and square-shaped blocks or cutouts for the students.

Give students a square attribute block. Have the students trace the edges of the block and identify sides and corners. Then ask,

- **How many sides does this block have?**
- **How many corners?**
- **This block has a square shape. All square shapes have four sides and four corners, and all the sides are the same size. All the corners are the same size.**
- **What does every square shape have?**

Give students a circular attribute block, and repeat the procedure above. Then make up riddles about squares and circles for students to identify. Use riddles similar to the following:

- **I do not have legs or arms. People sit on me. Furniture sits on me. I am made with many square shapes. What am I? Am I a wall or the floor?**

After students name the object, ask someone to find it.

Lesson Follow-Up

If students have difficulty completing this activity, have them complete Activities 20 and 21.

Conceptual Development Activities

Activity 19

Developing **Objective**

Students identify spatial relationships such as *on, off, up,* and *down*.

Display a table and a smaller object, such as a toy truck. Put the truck on the table, and tell the students,

- **The truck is on the table. Where is the truck?**

Take the truck off the table.

- **The truck is off the table. Where is the truck?**

Then put the truck on and off several items, each time asking,

- **Where is the truck?**

Give individuals instructions involving *on* and *off*. For example, say,

- **Put your finger on your nose. Take your finger off your nose.**
- **Put a block on the table. Take the block off the table.**

Next, demonstrate *up* and *down*. As you hold an object, raise and lower your arm, saying,

- **I am holding this truck up. Where is the truck?**
- **I am holding the truck down. Where is the truck?**

Give individuals instructions involving *up* and *down*. For example, say,

- **Raise your thumb up. Put your thumb down.**
- **Lift one leg up. Put your leg down.**

Lesson Follow-Up

If students have difficulty completing this activity, have them complete Activity 22.

Activity 20

Emerging **Objective**

Students recognize a common object with a two-dimensional shape.

Display a familiar classroom object that is essentially two-dimensional, such as a sheet of paper, a round paper coaster, or a pattern block. Show an object, such as the sheet of paper, and two other objects, such as a clock or a book, and say,

- **Show me/Tell me which one is shaped like the paper.**

Compare the sheet of paper to the object the student selects. Ask the student to find the shape of the found object and trace it with a finger. If the student is correct, confirm it by saying something like "You are correct. This paper has a rectangular shape, and this book has a rectangular shape."

If the student is incorrect, model as you trace the edge of the paper. "Look at the shape of the paper. Straight line, straight line, straight line, straight line. Four straight lines touching." Then help the student find an object with a rectangular shape.

Lesson Follow-Up

If students have difficulty completing this activity, have them reinforce their skills by completing ***Building Blocks*** Matching Shapes, which focuses on matching shapes to real-world objects.

Activity 21

Emerging **Objective**

Students recognize a common three-dimensional object.

Display two familiar objects on a table. The items should have an identifiable shape, such as a ball, a box, a book, a can, an orange, a drinking glass, and a cone.

Give an object to a student, and say,

- **Look carefully. Show me the ball.**

If necessary, give the student choices: Ask "Is this an orange or a can?" or rephrase to a *yes* or *no* question. If the student responds incorrectly, name the item in a complete sentence.

Repeat the process until all students have had a chance to identify an object. Then point to an object, and ask the group to name it. If necessary, name the item again in a complete sentence.

Lesson Follow-Up

If students have difficulty completing this activity, create two sets of matching picture cards. Lay one set of cards faceup, and choose a card from the second set. Say, "Here is a picture of an orange. Show me/Tell me where you see another orange." Repeat with other pictures.

Activity 22

Emerging **Objective**

Students recognize movement that reflects a spatial relationship such as *up* and *down*.

Demonstrate *up* and *down*. As you hold an object, raise and lower your arm, saying,

- **I am holding this truck up. Is the truck up?**
- **I am holding the truck down. Is the truck down?**

Next, perform various movements that demonstrate *up* and *down*, and ask students to identify them. For example, raise your arms in the air and ask,

- **Are my arms up?**

Other movements could include standing up or sitting down, raising your hand or putting your hand in your lap, and raising or lowering your foot.

Lesson Follow-Up

If students have difficulty completing this activity, use position words purposefully throughout the day.

Conceptual Development Activities

Activity 23

Independent **Objective**

Students use terms such as *big, small, long,* and *short* to compare overall size and length of objects.

Have available pairs of objects of very different sizes for comparison, such as pencils, crayons, trucks, cars, balls, and plastic containers.

Show any two objects of very different sizes. Ask,

- **Tell me which one is big. Which one is small?**
- **Tell me which one is long. Which one is short?**

Encourage students to use the descriptive words when they answer. Continue showing pairs of objects and asking the questions. You may combine objects in a variety of ways so that students notice indirectly that size is relative: a toy truck might be big when compared to a marble, but it might be small when compared to a table. A pencil might be long compared to a crayon, but it is short compared to a ball bat.

Lesson Follow-Up

If students have difficulty completing this activity, have them complete Activity 24.

Activity 24

Developing **Objective**

Students use terms such as *big* and *little* to identify the size of objects.

Have available several pairs of objects of very different sizes. Show one pair next to each other. Touch the appropriate object as you say

- **These are not the same size. This (object) is big. This (object) is little. Which is big? Which is little?**

Compare several more pairs using the same procedure. Then ask students to choose two objects and tell which is big and which is little. Ask one student to find something big in the classroom. Then ask another student to find something little. Encourage students to use the descriptive words when they answer.

Lesson Follow-Up

If students have difficulty completing this activity, have them complete Activity 25.

Conceptual Development Activities

Activity 25

Emerging **Objective**

Students recognize differences in size of objects.

Have available several objects in a variety of sizes. Choose two objects that are of vastly different sizes. Ask,

- **Are these things the same size? Show/Tell me which one is bigger.**

Encourage students to handle the objects and compare them. If necessary, put one in front of the other, one on top of the other, or one inside the other, and ask questions such as

- **Show me/Tell me which one is bigger.**
- **Show me/Tell me which one is smaller.**

Repeat comparisons with additional objects.

Lesson Follow-Up

If students have difficulty completing this activity, use objects that are alike in every way except size. This will help students focus only on size. Have them compare things like pencils, red crayons, sheets of paper, and plates. They can manipulate these pairs to see whether one fits exactly on top of the other or one hides all of the other.

Activity 26

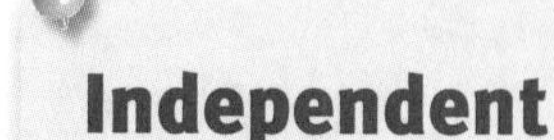

Independent **Objective**

Students match two-element repeating patterns of sounds, physical movements, and objects.

Begin with a very simple sound pattern, such as tapping four times on a desk or on the floor.

- **A pattern is something that happens over and over. Listen. I will tap a pattern.**

Tap four times.

- **You do it.**

If necessary, repeat the pattern one time.

Continue making simple sound patterns, that repeat at least four times. You may vary the rhythm, but do not use more than two different sounds in your pattern.

Next, create a very simple movement pattern, such as *clap, stomp* four times, and repeat the basic procedure above. Continue making simple movement patterns that repeat at least four times. You may vary the rhythm, but do not use more than two different movements in your pattern.

Finally, create a simple object pattern with pattern blocks or color cubes, using the basic procedure above. Continue making simple patterns that repeat at least four times. Do not use more than two items in your patterns. Invite students to create patterns for others to follow.

Lesson Follow-Up

If students have difficulty completing this activity, have them complete Activity 27.

Conceptual Development Activities

Activity 27

Developing **Objective**

Students match identical sounds, physical movements, and objects.

Tell students,

- **Listen. I will make two sounds. You will tell me whether they are the same.**

Tap two times.

- **Were the sounds the same?**

If necessary, repeat one time. Continue making simple sound pairs, such as two finger snaps, foot tap and clap, two claps, a tongue click and finger snap.

Next, demonstrate two identical movements. Tell students,

- **I am going to do two movements with my hands. You tell me whether the movements are the same.**

Begin with closed fists. Open one fist. Hold it open. Open the other fist. Hold it open.

- **Were the movements the same?**

Continue making pairs of simple movements. Vary the combinations so that sometimes the movements are the same and sometimes they are different.

Finally, repeat the activity with pairs of objects, such as color cubes.

If students have trouble, demonstrate again, or ask individuals leading questions. Continue making cube combinations. Vary the combinations so that sometimes the pairs are identical and sometimes they are different.

Lesson Follow-Up

If students have difficulty completing this activity, have them complete Activity 28.

Activity 28

Emerging **Objective**

Students recognize two objects that are identical to each other.

Have available several pairs of common classroom objects that are as nearly identical as possible, such as new pencils, new crayons, board erasers, blocks, and cubes.

Show a familiar object. Ask,

- **Is this a (object)? Yes, this is a (object).**

Show the matching object. Place the objects in front of the students along with another (non-matching) object, and ask students to look at the objects carefully. Then ask:

- **Show me/Tell me which one is the same. Show me/Tell me which one different.**

If necessary, help students identify matching characteristics. Repeat the procedure with several more pairs.

Lesson Follow-Up

If students are having difficulty, have them compare two items that differ in only one attribute, such as color or size. Make sure the difference is as clear as possible, such as a dark color and a light color or a large box and a small box.

Conceptual Development Activities

Activity 29

Independent **Objective**

Students relate activities to a time period by identifying concepts such as day, night, morning, and afternoon.

Discuss activities students do at different times of day.

- **What do you do in the morning before you come to school?**

If necessary, lead with more specific questions, such as "Do you brush your teeth in the morning? Do you get dressed in the morning? Do we read books in the morning? Do you go to learning centers in the morning?"

- **What other things do we do in the morning?**

Follow similar lines of questioning for afternoon, night, and day.

Lesson Follow-Up

If students have difficulty completing this activity, have them complete Activity 30.

Activity 30

Developing **Objective**

Students relate daily events to a time period by identifying concepts such as day and night.

Show photographs that represent day and night, and have a general discussion about day and night. Then discuss activities students do during the day and at night. Use questions such as the following, that are appropriate to the students' experiences:

- **Do you get ready for school in the daytime?**
- **When do you ride your bike, during the day or at night?**
- **When do you go to school, during the day or at night?**
- **When do you get ready for bed?**
- **When do you sleep for a long time?**
- **What other things do you do during the day?**
- **What other things do you do at night?**

Lesson Follow-Up

If students have difficulty completing this activity, have them complete Activity 31.

Activity 31

Emerging **Objective**

Students recognize common activities that occur every day.

Use photographs of a variety of daily activities and/or objects that represent the activities, such as flatware or plates for eating, a drinking glass, a toothbrush, soap and towel, shoes, a T-shirt, jeans, pajamas, crayons, and books, and photographs or objects that represent other, less frequent activities, such as zoo animals, doctors, and snow. Show students an object or a photograph from each category. Ask questions such as

- **Show me/Tell me which one you see every day.**

If necessary, name the object with a complete sentence or suggest an activity, such as

- **This is a book. We read books every day. Show me how to hold a book.**

Continue with a variety of familiar objects and activities.

If you use cues to activities in the classroom, review them.

- **What do I do when I want you to listen?**
- **What do I do to tell you it's time to (read a book)?**

Lesson Follow-Up

If students are not able to identify or name common activities, be very specific with cues. For example, if you flash the lights to get the group's attention, explain or confirm the procedure until it becomes automatic for students. "I flash the lights like this so you will stop what you're doing and listen to me. When I flash the lights, you stop what you are doing and listen to me."

NOTES

Conceptual Development Activities

Activity 1

Independent Objective

Students identify the meaning of addition as adding to and subtraction as taking away from.

Have available several small classroom objects, such as counters, blocks, or cars, and adjust the activity. Set one object in front of you, and say:

- **Here is one counter. It is a set of one.**
- **Add one to make a set of two.**
- **What did you do?**
- **How many in all?**
- **Take one away to make a set of one.**
- **What did you do?**
- **Now how many in all?**

If necessary, demonstrate the process. Repeat the activity with various quantities from 1 to 3. Then give each student three objects so students can work individually. Repeat the procedure above.

Lesson Follow-Up

If students have difficulty completing this activity, have them complete Activity 3.

Activity 2

Independent Objective

Students use counting and one-to-one correspondence as strategies to solve addition facts with sums to 10 and related subtraction facts represented by numerals with sets of objects and pictures.

Draw five counters on the board. Box the counters, and ask:

- **How many are in this set?**

If students don't recognize five, ask them to count the set. Write 1, 2, 3, 4, 5 below the counters, and 5 below the set. Draw a set of three counters. Box the counters, and ask:

- **How many are in this set?**

If students don't recognize three, ask them to count the set. Write 1, 2, 3 below the counters, and 3 below the set.

- **Let's add these sets together. How can we find out how many counters in all?**

Students should count each counter as you point to it. Write the numbers below the counters, and 8 below the sets. Repeat with other combinations. Draw seven counters on the board. Box the counters. Ask:

- **How many are in this set?**

Write the numbers below the counters, and 7 below the set.

- **Let's make two sets. I will take some away.**

Box 2 counters.

- **I took away 2.**

Write the numbers below the counters, and write 2 below the set.

- **How many are left?**

Write the numbers below the counters, and 5 below the set. Repeat with other combinations.

Lesson Follow-Up

If students have difficulty completing this activity, have them complete Activity 4.

Conceptual Development Activities

Activity 3

Developing Objective

Students demonstrate understanding of the meaning of joining (putting together) and separating (taking apart) sets of objects.

Have available several small classroom objects, such as counters, blocks, or cars.

Place one object in front of you.

- **Here is one counter. I made a set of one.**

Place a second object in front of you, away from the first counter.

- **Here is one counter. I made a set of one. Now I am going to join the sets.**

Push the two counters together.

- **I put a set of one with a set of one. Now I have a new set. How many in all?**

Give each student three objects. Direct the students through the procedure above. Repeat the activity with 1 add 2 and 2 add 1.

Next, make a set of two in front of you.

- **Here are two counters. I made a set of two. Now I am going to take this set apart and make new sets.**

Separate the two objects by several inches.

- **I separated a set of two into two different sets. How many in each new set?**

Direct the students through the procedure above. Repeat the activity with 3 take away 2 and 3 take away 1.

Lesson Follow-Up

If students have difficulty completing this activity, have them complete Activity 5.

Activity 4

Developing Objective

Students use one-to-one correspondence as a strategy for solving simple number stories involving joining (putting together) and separating (taking apart) with sets of objects to 5.

Have available five blocks and five dolls. Set two dolls in front of you, and say:

- **How many dolls in this set?**

If students don't recognize two, help them count.

- **I want to give a block to each doll. How can I do that?**

Help students match one-to-one and count to make sure each set has the same amount.

Indicate the combined set.

- **How many in all?**

If students don't recognize four, help them count on.

Help students form sets and count on as you repeat the activity with a variety of sets of dolls and blocks up to a total of five blocks.

Next, show four dolls.

- **How many dolls in this set?**

If students don't recognize four, help them count.

- **I'm going to put away some of the dolls.**

Set three dolls to the side.

- **How many dolls did I take away?**

If necessary, help students count the dolls.

- **How many are left?**

Help students form sets and take away as you repeat the activity with combinations up to a total of 5.

Lesson Follow-Up

If students have difficulty completing this activity, have them complete Activity 6.

Conceptual Development Activities

Activity 5

Emerging **Objective**

Students recognize when an object or person is added to (addition) or is taken away from (subtraction) a situation.

Have available several small classroom objects, such as counters, blocks, or cars.

Place two objects in front of you.

- **Here are some counters. We're going to play a game. While you cover your eyes, I will do something to this set. You will tell me what I did. Cover your eyes.**

Remove one object.

- **Open your eyes. Look at the set. What is different? Did I add something to the set? Did I take something from the set?**

To confirm, show and describe what you did.

Repeat with a variety of sets. Initially, remove enough objects so that the difference is obvious.

Place two objects in front of you, and tell the students to cover their eyes. This time take an object from the set.

- **Open your eyes. Look at the set. What is different? Did I take something from the set? Did I add something to the set?**

To confirm, show and describe what you did.

Repeat with a variety of sets. Initially, remove or add enough objects so that the difference is obvious.

Lesson Follow-Up

If students have difficulty completing this activity, work with people instead of objects, and limit the sets to 3 and below.

Activity 6

Emerging **Objective**

Students solve problems involving small quantities of objects or actions using language, such as *enough, too much,* or *more.*

Use counters and familiar classroom objects for this activity.

Show one counter, and say:

- **Here is one counter. I need a counter for each of us—two counters. Do I have enough?**

To correct, show your counter again, and say:

- **Here is one counter. I need two counters. I need one counter for you and one counter for me. Do I have enough counters? Two is more than one. I need one more. Please give me/show me one more counter.**

Repeat the activity with combinations of objects up to 3.

Next, give a student a counter, and say:

- **Here is a counter. Show/Tell me how many counters you have.**

Next, show two counters as you say:

- **Here are two counters. I need only one. I have too much. I will take away one counter.**

Give the student two counters, and say:

- **Here are two counters—one for each of us.**
- **You need one counter. Do you have too much? Show/Tell me what you should do.**

Repeat the procedure with other small quantities.

Lesson Follow-Up

If students have difficulty completing this activity, continue working on comparing sets of one, two, and three, using one-to-one correspondence. Help the students identify which set has more or fewer and how they know.

Conceptual Development Activities

Activity 7

Independent **Objective**

Students compare and order numbers 1 to 10.

Have available sets of common classroom objects, such as counters, toys, or blocks, for each student.

Tell the students to make a set of two counters.

- **Now make another set with the same amount.**
- **How do you know you have the same amount?**

If students suggest matching the sets or counting, ask a volunteer to do it. Ask students to count aloud the counters in each set, making sure students count in order.

- **Put all your counters together. Now we have a set of four.**
- **Now make another set with more than four counters.**
- **How do you know the set has more than four counters?**

If students suggest matching the sets or counting, ask a volunteer to do it. Ask students to count aloud the counters in the set, making sure students count in order.

- **Put all your counters together. Let's make a new set.**
- **Make a set with five counters.**
- **Now make another set with fewer than five counters.**
- **How do you know the set has fewer than five counters?**

If students suggest matching the sets or counting, ask a volunteer to do it. Ask students to count aloud the counters in the set, making sure students count in order.

Repeat the activity with a variety of sets.

Lesson Follow-Up

If students have difficulty completing this activity, have them complete Activity 10.

Activity 8

Independent **Objective**

Students use one-to-one correspondence to count sets of objects or pictures to 10.

Prepare math mats for the students. Tape together a sheet of red paper and a sheet of yellow paper to form a mat with red on the left and yellow on the right. Give a set of 20 counters and a divided math mat to each student. Demonstrate as necessary.

- **Show me one counter. Put one counter on the red/left side of your mat.**
- **Put one counter on the yellow/right side. Now you have two sets. How many counters are in each set?**
- **Put another counter on the red side. You still have two sets. Which side has more counters? How many counters are in each set?**

Have the students remove all the counters. Give directions for making additional sets through 10.

Lesson Follow-Up

If students have difficulty completing this activity, have them complete Activity 11.

Activity 9

Independent Objective

Students represent numbers to 10 using sets of objects and pictures, number names, and numerals.

Have available objects and pictures of objects in sets of 1 through 10 and sets of 10 counters or blocks for each student.

Show one of the pictures or objects, and ask the students to count the objects with you. Point to each object as you count. Continue showing pictures and counting objects.

Give a set of 10 counters to each student. Tell the students:

- **We're going to play a game. I will say a number, and you make a set with that many counters.**

Say numbers randomly, and help students check their sets by counting.

Then tell the students:

- **We're going to change the game. I will write a numeral on the board. I will read it and then you will say it. Then you will make a set with that many counters.**

Change the game again by writing the numeral on the board but not saying the number word or word for the numeral.

Lesson Follow-Up

If students have difficulty completing this activity, have them complete Activity 12.

Activity 10

Developing Objective

Students use one-to-one correspondence to compare sets of objects to 5.

Show two counters in each hand, and ask,

- **Do I have the same amount of counters in each hand?**
- **How do you know?**

If students suggest matching the sets, ask a volunteer to do it. If not, show students how to match the sets one-to-one. Then ask:

- **Does each set have the same amount?**

Show two more sets, one with several more than the other, and ask:

- **Does each set have the same amount?**
- **How do you know?**

Again, use one-to-one correspondence to check the answer. Ask questions such as

- **Show me/Tell me which set has more.** (Point to the larger set.) **This set has more than** (point to the smaller set) **this set.**
- **Show me/Tell me which set has less.** (Point to the smaller set.) **This set has less than** (point to the larger set) **this set.**

Repeat the activity with various objects up to 5.

Lesson Follow-Up

If students have difficulty completing this activity, have them complete Activity 13.

Conceptual Development Activities

Activity 11

Developing Objective

Students use one-to-one correspondence to count sets of objects to 5 arranged in a row.

Give a set of 5 counters to each student. Have available a set of five toys or books for yourself.

Show a toy, and ask:

- **How many toys do I have?**
- **Show me a set of one counter.**
- **Did you show me one counter?**

If students suggest matching the sets, ask a volunteer to do it. If not, show students how to match the sets one-to-one and then count. Then show students 2 toys in a row, and ask:

- **How many toys do I have?**
- **Show me a set of two counters.**
- **Did you show one counter for each toy? How do you know?**
- **Does each set have the same amount?**

Have a volunteer match counters to toys. Then have the students count their counters together. Continue with the same procedure until the students have counted sets to 5.

Lesson Follow-Up

If students have difficulty completing this activity, have them complete Activity 13.

Activity 12

Developing Objective

Students represent quantities to 5 using sets of objects and number names.

Have available objects or pictures of objects in sets of 1 through 5 and sets of 5 counters or blocks for each student.

Show one of the pictures or objects, and ask students to count the objects with you. Point to each object as you count. Continue showing pictures and counting objects.

Give a set of 5 counters to each student. Tell the students:

- **We're going to play a game. I will say a number, and you make a set with that many counters.**

Say numbers randomly, and help students check their sets by counting.

Lesson Follow-Up

If students have difficulty completing this activity, have them complete Activity 14.

Conceptual Development Activities

Activity 13

Emerging **Objective**

Students associate quantities with language, such as *many, a lot,* or *a little.*

Fill a box with counters. Tell the students:

- **I have a bowl filled with counters. There are a lot of counters in this bowl.**

If students are able, ask a student to take some counters and put them on the table.

- **You took some counters from the bowl. Did you take a lot of counters or a little bit?**

Point to each set as you say:

- **Are there more counters in the bowl or on the table?**
- **Are there as many counters on the table as there are in the bowl?**

Have the student put the counters into the bowl and repeat the activity.

Continue by asking:

- **Show me/Tell me about something we have a lot of in our classroom.**

If necessary, point out things such as crayons, paper, blocks, and books. Use language to describe how many: **We have a lot of crayons**.

- **Show me/Tell me about something we have a little of in our classroom.**

Some things could be clay, plants, or television.

Lesson Follow-Up

If students have difficulty completing this activity, create special math boxes. Use two flat boxes, one much smaller than the other. Tape or staple the boxes together side by side. Show the student two piles of counters. Have them put the large set of counters into the big box and the small set of counters in the small box.

Activity 14

Emerging **Objective**

Students recognize rote counting 1 to 3.

Tell the students that you are going to play a counting game.

- **I will tell you something to do, but you must not start until I have counted 1, 2, 3.**
- **Let's practice. I will do it with you. After I say 3, touch your chin. 1. 2. 3.**

Do the activity with the students. If some children are having difficulty, ask a volunteer who is doing the activity correctly to help you demonstrate. An assitive technology device or an adult can assist a student.

- **Let's practice again. I will tell you something to do, but you must not start until I have counted 1, 2, 3. After I say 3, touch your nose. 1. 2. 3.**

Continue giving instructions that are appropriate for your students. Do the activity with them.

When students are able to do the activity with you, give the instruction and count, but let students do the activity on their own.

Lesson Follow-Up

If students have difficulty completing this activity, remind them to wait. For example, you might say, "1. Wait. 2. Wait. 3. Go."

Conceptual Development Activities

Activity 15

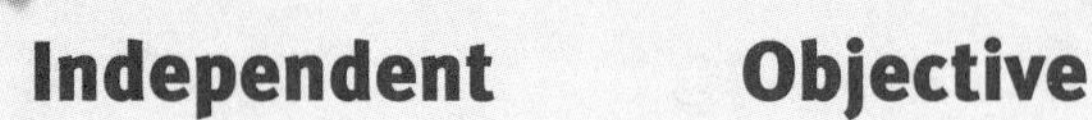

Independent Objective

Students sort and describe two dimensional shapes by single attributes, such as number of sides and straight or round sides.

Give each student or pair of students a pile of circular, square, and triangular attribute blocks. Ask the students to sort the blocks into groups. If the students sort by shape, say:

- **Tell me about your groups.**
- **Which group has curved edges? Do those blocks have corners? Which group(s) has straight sides?**

If the students sort into three- and four-sided groups, say:

- **Which group has four sides that are the same length? Do those blocks have corners? How many?**
- **Which has three sides? Do those blocks have corners? How many?**

If the students did not sort by shape, say:

- **Tell me about your groups.**
- **Let's try to make some new groups. Find all the blocks that have curved edges.**
- **What blocks do you have left?**
- **Put those blocks into groups.**
- **Tell me about your groups.**
- **Which group has four sides that are the same length? Do those blocks have corners? How many?**
- **Which has three sides? Do those blocks have corners? How many?**

Lesson Follow-Up

If students have difficulty completing this activity, have them complete Activity 17.

Activity 16

Independent Objective

Students combine two shapes to make another shape and identify the whole-part relationship.

Have available several pairs of cutouts or attribute blocks in a variety of shapes, including identical right triangles.

Give each student two cutouts or pattern blocks that are identical right triangles, and tell them:

- **Today you are going to work with a type of puzzle. Put the two triangles together to make another shape.**

When the students have constructed their new shapes, ask them to tell about their new shapes. Accept any shapes they create, and discuss sides and corners.

Then give the students additional cutouts or pattern blocks of a variety of shapes, and encourage students to make additional shapes.

Lesson Follow-Up

If students have difficulty completing this activity, have them complete Activity 18.

Conceptual Development Activities

Activity 17

Developing **Objective**

Students match and name common two-dimensional objects by shape, including square and circle.

Have available common objects with distinctive shapes, such as coins, plates, carpet tiles, sheets of paper, round paper coasters, or attribute blocks.

Give each student one square, one triangular, and one circular attribute block or cutout. Ask:

- **Show me/Tell me how they are alike.**
- **Show me/Tell me how they are different.**

If necessary, have the students trace the edges of the cutouts and identify sides and corners. They should notice that one block does not have corners.

Have students take turns holding up an attribute block so the other students can name its shape.

Show an object, such as the sheet of paper, and say:

- **Show me/Tell me something in the room that has this shape.**

Compare the sheet of paper to the object the student selects. If the student is correct, confirm it by saying something like "You are correct. This paper has a square shape, and this carpet tile has a square shape."

If the student is incorrect, model as you trace the edge of the paper, saying, "Look at the shape of the paper. Straight line, straight line, straight line, straight line. Four straight lines touching." Then help the student find an object with a square shape.

Lesson Follow-Up

If students have difficulty completing this activity, have them complete Activity 19.

Activity 18

Developing **Objective**

Students sort common two- and three-dimensional objects by size, including *big* and *little*.

Have available pairs of objects of very different sizes for comparison, such as pencils, crayons, trucks, cars, balls, and plastic containers.

Show any two objects of very different sizes. Ask:

- **Show me/Tell me which is big. Show me/Tell me which is little.**

Continue showing pairs of objects and asking the questions. You may combine objects in a variety of ways so that students notice indirectly that size is relative: a toy truck might be big when compared to a marble, but it might be small when compared to a table. A pencil might be long compared to a crayon, but it is short compared to a baseball bat.

Give pairs of students several objects for comparison. Encourage students to tell the group which things they think are big and which are little, and how they decided.

Lesson Follow-Up

If students have difficulty completing this activity, have them complete Activity 20.

Conceptual Development Activities

Activity 19

Emerging **Objective**

Students recognize common objects with two-dimensional shapes, such as circle or square.

Before the lesson, make sure there are several objects with a square shape and others with a circular shape scattered throughout the room. Give students a square pattern block or cutout. Have the students trace the edges of the block and identify sides and corners. If necessary, run your hands along the sides and corners, saying: "This is a side. This is a corner." When you have traced all four sides and corners, ask:

- **Show me/Tell me how many sides this block has. Show me/Tell me how many corners.**
- **This block has a square shape. All square shapes have four sides and four corners. All the sides are the same size. All the corners are the same size.**

Go on a "square shape hunt" with the students. Stop occasionally, and ask students to point to an object with a square shape.

Repeat the activity with a circular shape.

Lesson Follow-Up

If students have difficulty completing this activity, have them reinforce their skills by completing ***Building Blocks*** Geometry Snapshots, which focuses on matching shapes.

Activity 20

Emerging **Objective**

Students recognize common three-dimensional objects, such as balls (spheres) or blocks (cubes).

Display several common objects that are shaped like a cube or a sphere, such as balls, oranges, marbles, boxes, blocks, number cubes, or cube-shaped tissue boxes.

Give an object to a student, and say:

- **Look at this carefully. What is it?**

If necessary, name the item in a complete sentence; ask the student to repeat the sentence; and ask the group to repeat the sentence.

Repeat the process until all students have had a chance to name an object. Then point to an object, and ask the group to name it. If necessary, name the item again in a complete sentence.

Hold up a cube, and tell students:

- **Find something shaped like this.**

Repeat the activity with various objects, giving each students the opportunity to find a matching object.

Lesson Follow-Up

If students have difficulty completing this activity, have them reinforce their skills by completing ***Building Blocks*** Memory Geometry, which focuses on matching shapes.

Conceptual Development Activities

Activity 21

Independent **Objective**

Students match a two-element repeating visual pattern.

Have available several color cubes in two colors, red and black checkers, or two sets of toys, such as dinosaurs or teddy bears.

Tell the students:

- **I am going to make a pattern. A pattern is a model that is repeated. Here's my pattern.**

Describe the pattern as you make it:

- **One red cube. One blue cube. One red cube. One blue cube.**
- **One red and one blue cube are repeated. Let's say the pattern.**

Touch each cube as you say the pattern. After the final blue cube, ask:

- **What comes next in the pattern? How do you know?**

Repeat with another two-element visual pattern. After you've repeated the pattern once, ask the students to tell what comes next, continuing until you have repeated the pattern several times.

Give the students color cubes or checkers. Then give them instructions for creating and repeating a pattern.

Let volunteers lead the game while you and the students create and repeat the patterns.

Lesson Follow-Up

If students have difficulty completing this activity, have them complete Activity 22.

Activity 22

Developing **Objective**

Students match objects by single attributes, such as color, shape, or size.

Give each student a circular shape. Have them trace around the shape with a finger. Then ask them to tell you about the shape. Give each student several more circular shapes to examine and tell how they are alike and different. Have the students put the circular shapes to the side.

Follow the same procedure with triangular shapes. Then ask the students to put the two sets together and mix them up.

Hold up a circular shape, and say:

- **Find a piece that matches this.**

If students have difficulty, point to a circular piece, and discuss the characteristics. Have the student find pieces "like this one."

Repeat the procedure with triangular pieces.

Finally, have students put two pieces side by side or on top of each other and tell which is bigger and which is smaller.

Lesson Follow-Up

If students have difficulty completing this activity, have them complete Activity 23.

Conceptual Development Activities

Activity 23

Emerging **Objective**

Students recognize two objects that are the same size or color.

Display five or six pencils of two sizes that look alike except for size. Ask a student to choose two pencils that are the same size. If the student chose two pencils that are the same size, say:

- **Are they the same size?**
- **How do you know?**

If necessary, help the student place the pencils beside one another to show that they are the same length.

- **Let's try to find two pencils that are the same size. Put the pencils beside each other.**

Place the pencils side by side on a base line. Say:

- **Tell/Show me which pencils are long. Which pencils are short?**
- **Some of the pencils are long and some are short. Choose two long pencils. Now choose two short pencils.**

Repeat the activity, having students sort by color.

Lesson Follow-Up

If students have difficulty completing this activity, have them reinforce their skills by completing ***Building Blocks*** Memory Geometry, which focuses on matching size or color.

Activity 24

Independent **Objective**

Students measure length of objects using nonstandard units of measure and count the units.

Have available for each student several classroom objects they can use for measuring distance, such as blocks, large paper clips, new pencils, drinking straws, and craft sticks. Students will be measuring objects up to 10 units in length or width.

Demonstrate measuring an object, such as a carpet tile, with paper clips by laying paper clips end to end until you've covered one dimension of the tile. Then ask students to count the paper clips with you. Write the number on the board next to a drawing of a paper clip. Continue the activity with two additional tools that are longer or shorter than the paper clips.

Group the students into teams, and give each team a different measurement tool. Assign each team an object to measure. Give each team time to report their measurement; then record the measurements. Rotate the teams. Have the teams measure another team's object with their measurement tool. Record the measurements, and talk about why the numbers are different.

Lesson Follow-Up

If students have difficulty completing this activity, have them complete Activity 26.

Conceptual Development Activities

Activity 25

Independent **Objective**

Students compare objects by concepts of length—using terms, such as *longer, shorter,* and *same*—and capacity, using terms, such as *full* and *empty.*

For comparison, have available sets of three objects of different lengths and a fourth object that is the same as one of the first three, such as pencils, crayons, and toy trucks. Also, have available sand or other dry substance and several clear plastic containers for comparison of capacity.

Show two objects of different lengths. Ask:

- **Show me/Tell me which is longer. Show me/Tell me which is shorter.**

Add a third object to the group. Ask:

- **Show me/Tell me which is longest. Show me/Tell me which is shortest.**

Add the fourth object to the group. Ask:

- **Show me/Tell me which two are the same length.**

Continue showing sets of objects and asking the questions.

Next, show two empty plastic contains. Ask:

- **Are these containers empty or full?**

Fill one container with sand.

- **One container is empty. One container is full. Which container is full?**

Continue showing sets of containers and asking the questions.

Lesson Follow-Up

If students have difficulty completing this activity, have them complete Activity 27.

Activity 26

Developing **Objective**

Students measure length of objects using nonstandard units of measure.

Have available for each student several classroom objects they can use for measuring distance, such as blocks, pencils, drinking straws, and shoes. Students will be measuring objects up to 5 units in length or width.

Demonstrate measuring an object such as a book with blocks by laying blocks end to end until you've covered one dimension of the book. Then ask the students to count the blocks with you. Write the number on the board next to a drawing of a block.

Continue the activity with two additional tools that are longer or shorter than the blocks. Then ask volunteers to measure other objects with the blocks.

Group the students into teams, and give each team a different measurement tool. Assign each team an object to measure. Give each team time to report their measurement; then record the measurements.

Lesson Follow-Up

If students have difficulty completing this activity, have them complete Activity 28.

Conceptual Development Activities

Activity 27

Developing **Objective**

Students compare objects by length using terms, such as *long* and *short*.

Have available pairs of objects of very different lengths for comparison, such as pencils, crayons, trucks, cars, and shoes.

Show any pair of objects. Ask:

- **Show me/Tell me which is long. Show me/Tell me which is short.**

If necessary, show students how to compare objects for length.

Continue showing pairs of objects and asking questions. You may combine objects in a variety of ways so that students indirectly notice that size is relative.

Give sets of pencils and crayons to pairs of students. Have students make pairs of objects—one that is long and one that is short. The objects do not have to be of the same type.

Discuss the objects with the students. Touch any pair of objects. Ask:

- **Show me/Tell me which is long. Show me/Tell me which is short.**

Lesson Follow-Up

If students have difficulty completing this activity, have them complete Activity 28.

Activity 28

Emerging **Objective**

Students recognize similarities and differences in size of common objects.

Have available a variety of toys, crayons, paintbrushes, and other common objects in several sizes. Begin by showing the students the paintbrushes, and ask:

- **Are these paintbrushes the same size?**
- **Tell me about/Show me a big paintbrush.**
- **Tell me about/Show me a little paintbrush.**

Remove the paintbrushes and show some crayons, and ask:

- **Are these crayons the same size?**
- **Tell me about/Show me a big crayon.**
- **Tell me about/Show me a little crayon.**

Place a crayon near a paintbrush, and ask:

- **Are the crayon and the paintbrush the same size?**
- **Tell me about/Show me the one that is big.**
- **Tell me about/Show me the one that is little.**

Continue with additional objects.

Place up to three objects on the table. Ask a volunteer to find the thing that is biggest. Ask another volunteer to find the thing that is littlest. Help the student verify the answer by comparing objects.

Lesson Follow-Up

If students have difficulty completing this activity, have them reinforce their skills by completing ***Building Blocks*** Deep Sea Compare, which focuses on comparing sizes.

Conceptual Development Activities

Activity 29

Independent **Objective**

Students solve real-world problems involving addition facts with sums to 10 and related subtraction facts using numerals with sets of objects and pictures.

Display 2 blocks or counters.

- **This is how many counters I have. How many do I have?**
- **I need one more.**

Put one more with the set.

- **How many are in all?**

Give each student ten blocks or counters.

- **Make a set of one block. You have one block. You need two more. Show two more.**
- **How many did you begin with? How many did you put into the set?**
- **How many in all?**

Repeat with a variety of sets, up to a sum of 10, using the blocks.

Next display five blocks or counters.

- **How many?**

Take one away.

- **How many did I take away from the set?**
- **Now how many are there in all?**

Have students form a set with a specific number of counters. Then have them take away a specific number, and follow the same line of questioning.

- **How many?**
- **How many did you take away from the set?**
- **Now how many are there in all?**

Lesson Follow-Up

If students have difficulty completing this activity, have them complete Activity 30.

Activity 30

Developing **Objective**

Students solve real-world problems involving simple joining (putting together) and separating (taking apart) situations with sets of objects to 5.

Display 3 blocks or counters.

- **This is how many counters I have. How many?**

Count the counters with the students.

- **I need one more.**

Put one more with the set.

- **How many are in all?**

Count the counters with the students. Give each student five blocks or counters.

- **Make a set of one counter.**

Confirm the number of counters.

- **You have one counter. You need two more. Show me two more.**

Confirm the number of counters.

- **How many did you begin with? How many did you put into the set? How many in all?**

Repeat with a variety of sets, up to a sum of 5, using the blocks. Display five blocks or counters.

- **How many?**

Take one away.

- **How many did I take away from the set?**
- **Now how many are there in all?**

Have students form a set with a specific number of counters. Then have them take away a specific number, and follow the same line of questioning.

- **How many?**
- **How many did you take away from the set?**
- **Now how many are there in all?**

Lesson Follow-Up

If students have difficulty completing this activity, have them complete Activity 31.

Activity 31

Emerging **Objective**

Students solve simple problems involving putting together and taking apart small quantities of objects.

Display one block or counter.

- **Tell/Show me how many.**

Show another set of one block or counter.

- **Tell/Show me how many.**

Put the two sets together.

- **I put the sets together. Tell/Show me how many in all.**

Give each student two blocks or counters. Tell them to make or show a set of one.

- **Tell/Show me how many.**

Have them put another counter with it.

- **Tell/Show me how many you put into the set.**
- **Tell/Show me how many in all.**

Repeat several times with different combinations of counters up to a sum of 3.

After students are able to join the sets without counting each time, work on taking away from sets.

Display three blocks or counters.

- **Tell/Show me how many.**

Take one away.

- **Tell/Show me how many I took away from the set.**
- **Now tell/show me how many there are in all.**

Have students form a set with a specific number of counters. Then have them take away a specific number, and follow the same line of questioning.

Lesson Follow-Up

If students have difficulty completing this activity, have them reinforce their skills by completing ***Building Blocks*** Barkley's Bones, which focuses on simple problems.

NOTES

2

Conceptual Development Activities

Activity 1

Independent **Objective**

Students group objects into sets of tens and ones.

Give each student 20 math-link cubes.

- **Use your math-link cubes to make a group of one.**
- **Add one math-link cube to your group of one. How many math-link cubes are in the group?**

If necessary, stop periodically to count and verify the number of math-link cubes in the group. Continue until students have a group of ten.

- **You have a set, or a group, of ten. Making groups of ten helps us work faster and better.**
- **Keep your group of ten. Near it, make a new group of one.**
- **How many groups of ten do you have? How many groups of one?**
- **Let's count. We'll begin with the group of ten. We know there are ten in that group, so we can count on from ten. Touch each group as we count.**
- **Ten. Count on. Eleven. So, one group of 10 and one group of 1 is 11 in all.**

Continue until you've built two groups of 10. After you have repeated the activity above on several occasions and students can do it easily, have students make groups of 10 and 1 in random order.

Lesson Follow-Up

If students have difficulty completing this activity, have them complete Activity 5.

Activity 2

Independent **Objective**

Students match representations of numbers through 20 using pictures, sets of objects, and numerals.

You will need number cards with pictures of objects and numerals representing 1 through 20. Students will need math-link cubes and numeral cards for 1 through 20.

Give students the math-link cubes and numeral cards. Tell them:

- **I am going to show you a picture of a set of objects. Then you will use your math-link cubes to make the set.**

Show a small set, such as 3. Tell students:

- **This is set of three. Make a set of three.**
- **Count to make sure you have three. Show the card with the numeral 3.**

Continue randomly with sets up to 20.

When students are comfortable with the routine above, show a numeral, and ask the students to make the set. Encourage volunteers to pick a number card and lead the group.

Lesson Follow-Up

If students have difficulty completing this activity, have them complete Activity 6.

Conceptual Development Activities

Activity 3

Independent Objective

Students name ordinal numbers from first through fifth.

Make a row of five students, one at a time, from left to right:

- **We are going to make a row of students. (Maria), stand here. (Maria) is a set of one. (She/He) is the first person in our row.**
- **(Jon), stand here. (Jon) and (Maria) are a set of two. (Maria) is the first person in our row. Jon is second.**

Continue until you have a row of five. Then repeat with another group.

Ask a volunteer to lead the activity and make another group of five. Then have the students tell who is first, second, and so on.

Repeat the activity by having students arrange objects in order, first through fifth, on a sheet of paper. Once students have placed the objects on paper, have them draw circles around the objects and label them using index cards with the ordinal numbers first through fifth written on them.

Lesson Follow-Up

If students have difficulty completing this activity, have them complete Activity 5.

Activity 4

Independent Objective

Students create and compare sets of zero through twenty objects.

Have available 20 counters for each student and a large number line (at least 0–20) on the floor or on the board. Show an empty hand, and ask:

- **How many counters do I have in my hand? The set of counters in my hand is empty. The number that stands for none is 0. A set with none comes before a set with one.**

Show 1 counter in one hand and 2 counters in the other, and ask,

- **Do I have the same number of counters in each hand?**
- **How do you know?**
- **Which set has more?**
- **Which set has fewer?**

Give each student 20 counters. If students do not suggest matching the sets one-to-one, model the procedure for them. Model each step below, and then have the students do it. Each set should begin on the left and align with the set above it.

- **Make a set of one.**
- **Below the set of one, make a set of two.**
- **Which set has more?**
- **Which set has fewer?**
- **Show me/Tell me where we find 1 and 2 in the number line.**

Continue in the same way until students have made a set of 20.

Lesson Follow-Up

If students have difficulty completing this activity, have them complete Activity 5.

Conceptual Development Activities

Activity 5

Developing Objective

Students create and compare sets of up to five objects.

Show two counters in each hand, and ask,

- **Do I have the same number of counters in each hand?**
- **How do you know?**

If students suggest matching the sets or counting, ask a volunteer to do it. Then ask,

- **Does each set have the same amount?**

Show two more sets, one with several more than the other, and ask.

- **Does each set have the same amount?**
- **Which set has more?**
- **Which set has fewer?**
- **How many more would I add to make the same? Show me with counters.**

Give each student 15 counters. Model each step, and then have the students do it. Each set should begin on the left and align with the set above it, forming an array.

- **Make a set of one.**
- **Below the set of one, make a set of two.**
- **Which set has more?**
- **Which set has fewer?**
- **How many more would I add to make the same? Show me with counters.**

Continue in the same way until students have made a set of five.

Lesson Follow-Up

If students have difficulty completing this activity, have them complete Activity 7.

Activity 6

Developing Objective

Students match representations of numbers through 5 using pictures, sets of objects, and numerals.

You will need number cards with pictures of objects and numerals representing 1 through 5. Students will need counters and numeral cards for 1 through 5.

Give students the counters and the numeral cards. Tell them:

- **I am going to show you a picture of a set of objects. Then you will use your counters to make the same set you see in the pictures.**

Begin with one, and work up to five. Tell them:

- **This is set of one. Make a set of one.**
- **Count to make sure you have one. Show the card with the numeral 1.**

Continue in order up to 5.

Repeat the activity with sets in random order. When students are comfortable with the routine above, show a numeral, and ask the students to make the set. Encourage volunteers to pick a number card and lead the group.

Lesson Follow-Up

If students have difficulty completing this activity, have them complete Activity 8.

Conceptual Development Activities

Activity 7

Emerging **Objective**

Students use one-to-one correspondence to place objects in defined spaces.

Prepare math mats for you and the students. On each mat, make five blue circles a little larger than the blocks or counters you will use.

Show your mat, and say:

- **I have a math mat with blue spots on it, and I have some counters. I am going to match the spots and counters. Watch, and then you can do it.**

Set your counters on or near the mat. Move one counter to the first spot.

- **I am moving a counter to cover the first spot. I matched one counter to one spot.**

Continue until you've matched the spots and counters.

- **Now it's your turn.**

Give each student a mat and five counters.

- **Do it with me.**

Set your counters on or near the mat. Move one counter to the first spot.

- **I am moving a counter to cover the first spot. I matched one counter to one spot. Tell/show me what you did.**

Continue until you've matched the spots and counters, each time asking students to tell or show you what they did.

Lesson Follow-Up

If students have difficulty completing this activity, work with individuals in a learning center by placing small objects on a pegboard or in an egg carton.

Activity 8

Emerging **Objective**

Students count and create sets of objects.

Have available several large pictures of sets of 1 and 2 objects and sets of three counters or blocks for each student. You may use actual sets of objects if you prefer.

Show one of the pictures. Ask the students to count the objects with you. Point to each object as you count. Continue showing pictures and counting objects.

Give a set of 3 counters to each student. Tell the students:

- **We're going to play a game. I will say a number, and you make/show a set with that many counters.**

Say a number, and help students check their sets by counting.

Lesson Follow-Up

If students have difficulty manipulating the counters, say a number, and have students identify a picture that shows that number.

Conceptual Development Activities

Activity 9

Independent **Objective**

Students learn the meaning of the symbols +, −, and =.

Use a chalkboard, a whiteboard, or a magnetic board to draw or display one object, and say:

- **Here is one object. How many in this set?**

Write the numeral 1 below the set of one.

Write a plus sign to the right of the numeral. Touch the sign as you say:

- **This sign means to *add*. I am going to add one more.**

Add another object to the set.

- **I added one.**

Write the numeral 1 below the second set of one.

Read the number sentence as you touch each part.

- **One plus one.**
- **How many in all?**

Write the equal sign to the right of the second numeral. Touch each numeral and sign as you say:

- **This sign means *equals*. *Equals* means "the same as." We use it when we want to show how many in all. One plus one equals two.**

Give each student three counters, and have them work through the procedure with you.

Continue with one plus two and one plus three.

Repeat the procedure, beginning with two and taking away (subtracting) one.

Lesson Follow-Up

If students have difficulty completing this activity, have them complete Activity 12.

Activity 10

Independent **Objective**

Students solve problems using addition facts with sums through 10 and related subtraction facts.

Have available sets of pictures for adding and subtracting up to 10. You may prepare activity sheets for the students, or you may use a chalkboard for drawing or displaying sets, numerals, and algorithms.

Show a picture of a set of one, and say:

- **Here is a set. How many in this set?**

Have students count to confirm. Write the numeral 1 below the set of one. Write a plus sign to the right of the numeral. Touch the sign as you ask:

- **What does this sign mean?**

Add another object to the set.

- **I added one.**

Write the numeral 1 below the second set of one. Read the number sentence as you touch each part.

- **One plus one. How many in all?**

Write the equal sign to the right of the second numeral. Touch each numeral and sign as you say:

- **What does this sign mean? We use it when we want to show how many in all. One plus one equals two.**

Have the students count and use one-to-one correspondence to confirm various sums to 10. Repeat the procedure, beginning with two and taking away (subtracting) one.

Lesson Follow-Up

If students have difficulty completing this activity, have them complete Activity 13.

Conceptual Development Activities

Activity 11

Independent **Objective**

Students use objects to model and solve problems involving addition facts with sums through 10 and related subtraction facts.

Create addition and subtraction problems using familiar classroom objects. For example, using pennies, say:

- **Here is a set of two pennies. How many in this set?**
- **I need more pennies. Here are five more pennies. How many in this set?**
- **I am going to join the sets. A set of two plus a set of five. How many in all?**

Demonstrate as you say:

- **I have a set of five pennies.**

Have students count to confirm.

- **(Jo) wants to borrow two pennies, so I'll take two away from my set. I had five. I gave two to (Jo). How many do I have left?**

Have students demonstrate and count to confirm.

Repeat the procedure with various quantities up to sums of 10 using measurement and geometric shapes.

Lesson Follow-Up

If students have difficulty completing this activity, have them complete Activity 14. Expand the activity to other contexts. For example, have students measure the sides of a pattern block using nonstandard units and find the sum, measure the lengths of two objects and subtract the smaller number from the larger one, and so on.

Activity 12

Developing **Objective**

Students model problems involving combining sets and taking away from sets.

Have available several small classroom objects, such as counters, blocks, or cars.

Set one object in front of you.

- **Here is one block. I made a set of one.**

Set a second object in front of you, away from the first counter.

- **Here is one block. I made another set of one.**
- **Now I am going to join the sets.**

Push the two counters together.

- **I added a set of one to a set of one. Now I have a new set. How many in all?**

Give each student three objects. Direct the students through the procedure above.

Repeat the activity with other small quantities of objects.

Then, make a set of two in front of you.

- **Here are two counters. I made a set of two.**
- **Now I am going to take one from the set.**

Remove one object.

- **I had a set of two. I took away a set of one. How many are left?**

Direct the students through the procedure above.

Repeat the activity with other small quantities of objects.

Lesson Follow-Up

If students have difficulty completing this activity, have them complete Activity 15.

Conceptual Development Activities

Activity 13

Developing **Objective**

Students use sets of up to 5 objects to model addition and subtraction.

Have available sets of objects for adding and subtracting up to 5.

Show set of one, and say:

- **Here is a set. How many in this set?**

Have students count to confirm.

Add another object to the set.

- **I added one.**
- **One plus one. How many in all?**

Have the students count or use one-to-one correspondence to confirm. Continue additional problems up to sums of 5.

Repeat the procedure beginning with two and taking away (subtracting) one.

Lesson Follow-Up

If students have difficulty completing this activity, have them complete Activity 16.

Activity 14

Developing **Objective**

Students use sets of up to 5 objects to model problems involving joining and separating.

Create addition and subtraction problems using familiar classroom objects. For example, using blocks, say:

- **Here is a set of three blocks. How many in this set?**
- **I need more blocks to make a tower. Here are two more blocks. How many in this set?**
- **I am going to join the sets. A set of three plus a set of two. How many in all?**

Next, demonstrate as you say:

- **I have a set of four blocks.**

Have students count to confirm.

- **If I give (Kim) some blocks, (she/he) can make a tall tower, so I'll take two away from my set. I had four. I gave two to (Kim). How many do I have left?**

Have students demonstrate and count to confirm.

Repeat the procedure, beginning with additional sets up to sums of 5.

Lesson Follow-Up

If students have difficulty completing this activity, have them complete Activity 16.

Conceptual Development Activities

Activity 15

Emerging **Objective**

Students compare quantities through 3.

Use counters or other familiar classroom objects for this activity.

Show one counter, and say:

- **Here is one counter. I need two counters. Do I need more or less?**

Repeat, starting with one and adding two.

Give the student a counter, and say:

- **Here is a counter.**
- **Show/Tell me how many counters you have.**
- **You need two counters. Do you need more or less? Show me/Tell me how many more you need.**

Repeat the procedure, beginning with one counter and then adding two.

Show two counters as you say:

- **Here are two counters. I need only one. I need less. I will take away one counter.**

Give the student two counters, and say:

- **Here are some counters.**
- **Show/Tell me how many counters you have.**
- **You need one counter. Do you need more or less? Show me/Tell me what you should do.**

Repeat the procedure, starting with three and taking away two.

Lesson Follow-Up

If students have difficulty completing this activity, continue working on comparing sets of one, two, and three using one-to-one correspondence. Help the students identify which set has more or less and how they know.

Activity 16

Emerging **Objective**

Students use sets of up to 3 objects to solve problems involving joining and separating.

Create joining and separating problems using familiar classroom objects. For example, using a set of crayons, demonstrate as you say:

- **Here is a set of two crayons. How many in this set?**
- **I need more crayons. Here are three more crayons. How many in this set?**
- **I am going to join the sets. A set of two plus a set of three. How many in all?**

Have students count to confirm.

Demonstrate as you say:

- **I have a set of five crayons.**

Have students count to confirm.

- **(Terrell) wants to borrow one of my crayons, so I'll take one away from my set. I had five. I gave one to (Terrell). How many do I have left?**

Have students count to confirm.

Repeat the procedure with various small quantities.

Lesson Follow-Up

If students have difficulty completing this activity, have them repeat the activity using math-link cubes and smaller numbers.

Conceptual Development Activities

Activity 17

Independent Objective

Students measure objects using a ruler marked in inches.

Students will need a ruler divided into whole inches, no additional segments, beginning with 0 on the left end and ending with 12 on the right. You may make them or use commercial rulers. Students also need objects that are whole inches in length, such as books, crayons, and pencils.

Tell the students:

- **When we have measured things in the past, we may have measured with paper clips and crayons and pencils. These measuring items can vary in size and make it hard to get the same measurements every time. Now we have another tool for measuring. It is called a ruler. A ruler is divided into equal parts called inches.**

Distribute rulers, and let students examine them.

- **What have we used that looks like a ruler? The ruler is like a number line. It begins at 0 and goes up. Let's begin at 0 and count the inches.**

Demonstrate using a ruler to measure. Set the 0 on the ruler at the base of the object, and read to where the top of the object and a number on the ruler align.

Continue measuring objects. Then give objects to the students to measure.

Lesson Follow-Up

If students have difficulty completing this activity, have them complete Activity 20.

Activity 18

Independent Objective

Students compare and order objects by length.

For comparison, have available sets of three objects of different lengths, such as pencils, crayons, and toy trucks.

Show two objects of different lengths. Place them side by side. Ask:

- **Which is longer? Which is shorter?**

Add a third object to the group, in order from longest to shortest. Ask:

- **Which is longest? Which is shortest?**

Continue showing sets of objects and asking the questions.

Give each student a set of objects, and ask students to put the objects in order from longest to shortest. Continue with similar items, and then have students show longest to shortest with a combination of items, such as trucks and crayons.

Lesson Follow-Up

If students have difficulty completing this activity, have them complete Activity 21.

Conceptual Development Activities

Activity 19

Independent **Objective**

Students measure and compare lengths using a ruler divided into inches.

Students will need a ruler divided into whole inches, with no additional segments, beginning with 0 on the left end and ending with 12 on the right. Students will also need items to measure, such as crayons, drinking straws, and pieces of paper.

Tell the students:

- **We have some problems to solve today, and you can solve them with a ruler. Use your ruler to measure the longest part of your hand. How long is your hand? Is your hand as long as your foot? How can you find out? Measure your foot with your ruler.**
- **Remember, a ruler is a tool we use for measuring. We can find out how long an object is with a ruler. A ruler is divided into small parts called inches.**

Pose problems for the students to solve with their rulers. For example, if you have a box of crayon scraps, give each student a few crayons, and say:

- **I need a crayon that is about two inches long. Use your ruler to find one.**

Encourage students to pose problems for the group to solve.

Lesson Follow-Up

If students have difficulty completing this activity, have them complete Activity 21.

Activity 20

Developing **Objective**

Students measure lengths using nonstandard units.

Have available for each student several classroom objects they can use for measuring length, such as connecting cubes, large paper clips, new pencils, drinking straws, and craft sticks. Students will be measuring objects up to 10 units in length or width.

Demonstrate measuring an object such as a carpet tile with connecting cubes by laying the cubes end to end until you've covered one dimension of the tile. Then ask the students to count the connecting cubes with you. Write the number on the board next to a drawing of a cube.

Continue the activity with two additional tools that are longer or shorter than the connecting cubes.

Group the students into teams, and give each team a different measurement tool. Assign each team an object to measure. Give each team time to report their measurement; then record the measurements. Rotate the teams. Have the teams measure another team's object with their measurement tool. Record the measurements, and talk about why the numbers are different.

Lesson Follow-Up

If students have difficulty completing this activity, have them complete Activity 22.

Activity 21

Developing **Objective**

Students compare sizes of objects.

Have available a variety of objects of different lengths to use in solving real-world problems. For example, you could use three shoes of obviously different sizes; a small box, one object that will fit in the box, and two objects that won't; and a glove or mitten that will fit a child but not you.

Show a box and three objects, and ask:

- **Which object will fit in this box? How can we find out?**
- **(Linda), find another object you think will fit in this box.**

After the student has found an object, ask him/her to test it. Then ask other students to find things that will fit in the box and things that won't.

Lesson Follow-Up

If students have difficulty completing this activity, have them complete Activity 22.

Activity 22

Emerging **Objective**

Students classify objects as *long, short, big,* and *little.*

Have available pairs of objects of very different sizes for comparison, such as pencils, crayons, trucks, cars, and shoes.

Show any pair of objects. Ask:

- **Tell/Show me which is long. Which is short?**

If necessary, show students how to compare objects for length.

Continue showing pairs of objects and asking the questions.

Give sets of pencils and crayons to students. Have students make pairs of objects—one long object and one short object. The objects do not have to be of the same type.

Discuss the objects with the students. Touch any pair of objects. Ask:

- **Tell/Show me which is long. Which is short?**

Repeat the procedure, and compare objects that are *big* and *little.*

Lesson Follow-Up

If students have difficulty completing this activity, create cards with the terms on them, and have students use the cards to label objects on display.

Conceptual Development Activities

Activity 23

Independent Objective

Students create and extend repeating patterns with two elements.

Have available several color cubes in two colors, red and black checkers, or two sets of toys, such as dinosaurs or teddy bears.

Tell the students:

- **I am going to make a pattern. A pattern is a model that changes in a regular way. Here's my pattern that repeats over and over.**

Describe the pattern as you make it:

- **One red cube. One blue cube. One red cube. One blue cube.**
- **One red and one blue cube are repeated. Let's say the pattern.**

Touch each cube as you say the pattern. After the final blue cube, ask:

- **What comes next in the pattern? How do you know?**
- **What comes next in the pattern? How do you know?**

Repeat with another two-element visual pattern. After you've repeated the pattern once, ask the students to tell what comes next, continuing until you've repeated the pattern several times.

Give the students color cubes or checkers. Then give them instructions for creating and repeating a pattern. Let volunteers lead the game while you and the students create and repeat the patterns.

Lesson Follow-Up

If students have difficulty completing this activity, have them complete Activity 27.

Activity 24

Independent Objective

Students identify missing elements in repeating patterns with two elements.

Have available several items for creating two-element repeating visual patterns. You will also need a sheet of paper or poster board to shield the pattern when you remove part of it. Lay out a two-element pattern, such as yellow triangle, green square, yellow triangle, green square, yellow triangle, green square. Say:

- **This pattern has two parts that repeat. What is the pattern?**
- **Look at the pattern carefully. I am going to take away part of the pattern, and you will tell what I took away.**

Shield the pattern, and remove one element, such as the second yellow triangle. Show the pattern.

- **What is missing?**

If necessary, put the piece back, and say the pattern several times with the students. Then remove a different pattern piece for students to identify.

Repeat the procedure with additional two-element repeating patterns.

Lesson Follow-Up

If students have difficulty completing this activity, have them complete Activity 27.

Conceptual Development Activities

Activity 25

Independent Objective

Students compare the number of objects in sets of up to 20.

Have available counters, math-link cubes, blocks, and pictures of objects to make sets to 20.

Show two sets with the same number of objects.

- **Is this set of (4 objects) the same amount as this set of (4 objects)? Remember, when sets are equal, they have the same amount.**

Show two sets with obviously different amounts, such as 17 and 4, and ask:

- **Is this set of (17 objects) the same amount as this set of (4 objects)?**
- **The sets are not equal. They don't have the same amount.**

If students are not able to tell the difference, suggest matching one to one or counting.

Continue showing set pairs, using both objects and pictures, and have students identify whether they are equal.

Lesson Follow-Up

If students have difficulty completing this activity, have them complete Activity 29.

Activity 26

Independent Objective

Students apply rules involving adding 1 or 2.

Have available several objects for making sets to 10. Tell the students:

- **Here's a rule for a game we're going to play. You can start with any number. When you add one more, you get the next number.**

Demonstrate as you say:

- **Here is one toy. Here's one more. One toy and one more is . . . ?**

Repeat with two more numbers.

- **Here's another rule for the game. You can start with any number. When you add two more, you get the next number and the next number after that.**

Demonstrate as you say:

- **Here is one toy. Here are two more. One toy and two more are . . . ?**

Repeat with two more numbers.

It is difficult for some students to hold the idea of numbers in their mind. Practice with the students. If they are unable to keep the idea of the number and one more in their heads, continue using objects or counters.

- **We're going to play One More. First I will say a number. Then I will say "and one more." You will tell me the next number. I'll show you. The number is two and one more is . . . three.**

Repeat with two more numbers.

Repeat this game, going as high as possible with add "one more" and add "two more."

Lesson Follow-Up

If students have difficulty completing this activity, have them complete Activity 28.

Conceptual Development Activities

Activity 27

Developing **Objective**

Students extend repeating sound, movement, and object patterns with two elements.

For the third step in this activity, you will need several color cubes in two colors, red and black checkers, or two sets of toys, such as dinosaurs and teddy bears.

Tell students:

- **I am going to make a sound pattern. A pattern is a model that is repeated. Here's my sound pattern.**

Describe the pattern as you clap your hands and slap a table or your knees:

- **Clap, slap. Clap, slap. What's my sound pattern?**
- **Do it with me.**

Again, describe the pattern as you clap your hands and slap a table or your knees:

Continue with additional two-element sound patterns, and invite volunteers to lead the group.

Repeat the procedure with two-element movement patterns: flap your arms, scratch your head; shake your leg, shake your hand.

Repeat the procedure with two-element object patterns. Describe the pattern as you make it:

- **One red cube. One blue cube. One red cube. One blue cube.**
- **One red and one blue cube are repeated. Let's say the pattern.**

Touch each cube as you say the pattern. Then have the students say the pattern with you.

Continue with additional two-element object patterns: teddy bear, dinosaur; truck, car; crayon, pencil.

Lesson Follow-Up

If students have difficulty completing this activity, have them complete Activity 30.

Activity 28

Developing **Objective**

Students extend number patterns involving adding 1.

Have available several objects for counting to 5. Tell the students:

- **Here's a rule for a game we're going to play. You can start with any number. When you add one more, you get the next number.**

Demonstrate as you say:

- **Here are two toys. Here's one more. Two and one more is . . . three.**

Repeat with two more numbers. Tell students:

- **We're going to play One More. First I will say a number. Then I will say "and one more." You will tell me the next number. The number is one and one more is . . . ?**

Repeat with various numbers from 1 to 5. It's a big jump from being able to play the game with objects to being able to hold the idea of numbers in one's mind. Practice with the students. If they are unable keep the idea of the number and one more in their heads, continue using objects or counters.

Lesson Follow-Up

If students have difficulty completing this activity, have them complete Activity 31.

Conceptual Development Activities

Activity 29

Developing Objective

Students compare the number of objects in sets of up to 5.

Have available counters, blocks, and pictures of objects to make sets to 5.

Show two sets with the same number of objects.

- **Is this set of (4 objects) the same amount as this set of (4 objects)?**
- **How do you know?**

If necessary, suggest and/or demonstrate one-to-one correspondence.

Show two sets with obviously different amounts, such as 5 and 2, and ask:

- **Is this set of (5 objects) the same amount as this set of (2 objects)?**
- **How do you know?**

If necessary, suggest and/or demonstrate one-to-one correspondence.

Continue showing set pairs, using objects or pictures, and have students identify whether they are equal. Math-link cubes can be used to check work.

Lesson Follow-Up

If students have difficulty completing this activity, have them complete Activity 31.

Activity 30

Emerging Objective

Students identify sound patterns.

- **I will make two sounds. You tell/show me whether they are the same.**

Tap two times.

- **Were the sounds the same?**

If necessary, repeat.

Continue making simple sound pairs, such as two finger snaps; a foot tap and a clap; or a tongue click and a finger snap.

- **Now I am going to make a sound pattern. A pattern is a model that is repeated. Here's my sound pattern.**

Describe the pattern as you clap your hands and slap a table or your knees:

- **Clap, slap. Clap, slap. What's my sound pattern?**

Again, describe the pattern as you do it. Continue with two or three patterns. Then tell students:

- **I am going to make a sound pattern. Listen, and then tell/show me the pattern.**

Make a simple sound pattern, such as knock, tap, knock, tap. If the students are not able to recognize it, repeat it. Continue with two or three patterns.

Lesson Follow-Up

If students have difficulty completing this activity, have them complete Pattern Zoo from the ***Building Blocks*** program.

Conceptual Development Activities

Activity 31

Emerging **Objective**

Students use one-to-one correspondence to compare the number of objects in sets.

Show one counter in each hand, and ask,

- **Do I have the same amount of counters in each hand?**
- **How do you know?**

If students suggest matching the sets, ask a volunteer to do it. If not, show students how to match the sets one-to-one.

Show two counters in each hand, and ask,

- **Do I have the same amount of counters in each hand?**
- **How do you know?**

Then ask,

- **Does each set have the same amount?**

Show two more sets, one with several more than the other, and ask.

- **Does each set have the same amount?**
- **How do you know?**

Again, use one-to-one correspondence to check the answer.

Lesson Follow-Up

If students have difficulty completing this activity, have them complete Party Time from the ***Building Blocks*** program.

Activity 32

Independent **Objective**

Students combine shapes to complete puzzles.

Students will be completing shape puzzles. Use commercial pattern templates, or prepare templates based on the sizes and shapes of pattern blocks or shape cutouts you have. You may put several different templates on one sheet of paper. Have available several pairs of cutouts or attribute blocks in a variety of shapes, including identical right triangles and semicircles.

Give each student two cutouts or attribute blocks that are identical right triangles, and tell students:

- **Today you have some puzzles. You will put two parts together to make a whole new shape. Put the two triangles together to make a square. What two parts make a square? Where do these two pieces fit on your puzzle? Put the parts over the square on your paper.**

Give students another pair of shapes, and ask them:

- **Where do these parts fit in your puzzle? Put these two parts together to make another whole shape.**

Encourage students to tell about the shapes they created. Depending on the shape blocks/cutouts and templates you use, pairs will fit in more than one place. For example, two right triangles placed "side to side" will create another triangle. Two right triangles placed diagonal to diagonal will create a rectangle/square.

Lesson Follow-Up

If students have difficulty completing this activity, have them complete Activity 38.

Conceptual Development Activities

Activity 33

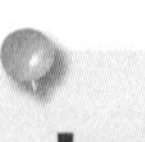

Independent Objective

Students use concepts of time including *first, next, before, after, yesterday, today,* and *tomorrow.*

Discuss with students things they do at school, relating them to time. For example:

- **What do you do first when you get to school?**

You might have to give hints and cues to clarify.

- **What do you do after you come into our room? What do you do next?**
- **What do we do after we read a book?**
- **What learning centers did you go to yesterday?**
- **What do we do before we have a snack?**
- **Will we have (art/music) today or tomorrow?**

Continue asking questions that help students sequence events.

Lesson Follow-Up

If students have difficulty completing this activity, have them complete Activity 39.

Activity 34

Independent Objective

Students identify days of the week on a calendar.

In the United States, most general calendars begin the week on Sunday; however, school and business calendars often show Monday as the first day of the week. Use a calendar that is appropriate for your situation.

Tell the students:

- **You know that days have names. Today is (Monday). What is tomorrow? What are the rest of the days' names?**

Display a current calendar that is large enough for students to see the day names and numbers, and explain:

- **We group days together into weeks. What are the names of the days of the week?**

Point out the days of the week.

- **We group weeks together into months. Now we are in the month of (October). We number the days of each month.**

Point out the numbers of the days.

- **Let's look at the days again. Every day on the calendar has a name. All the Mondays are in a column.**

Point out the Mondays. Then point to each Monday, and say, "This is Monday, (Month) (day)."

Ask different students to find the days of the week in order. Then name the days randomly, and ask the students to find them.

Lesson Follow-Up

If students have difficulty completing this activity, have them complete Activity 39.

Conceptual Development Activities

Activity 35

Independent Objective

Students describe, compare and contrast tools used for telling time.

Display a variety of analog and digital clocks and watches, and set up a problem for the students, such as:

- **We are going to have our snack a little later today. I don't want to forget it. What should I use to tell me when it's time for snack?**
- **These are all tools for telling time. Why do we want to know what time it is?**
- **Let's find out how these tools are alike and how they are different.**

Ask two students to each choose a tool for telling time, and then ask them to tell about their tool. When they have finished, ask the group:

- **How are these two tools alike? How are they different?**

Continue with other pairs for comparison.

Lesson Follow-Up

If students have difficulty completing this activity, have them complete Activity 39.

Activity 36

Independent Objective

Students describe uses of money.

Display a variety of bills and coins, and ask:

- **Have you ever seen any of these things? What are they?**
- **When we want something at a store, we give the store money. If I want a box of crackers, I trade the crackers for money. I use money to buy the crackers.**
- **Have you had money? What did you do with it?**
- **When do grown-ups use money?**

Show students a group of objects and coins. Have students point to the coins.

Lesson Follow-Up

If students have difficulty completing this activity, have them complete Activity 40.

Conceptual Development Activities

Activity 37

Independent **Objective**

Students compare objects by weight and capacity.

Have available a balance scale and objects of a variety of weights that will fit on the scale; for example, table-tennis balls and golf balls. For the capacity activity, have available containers in a variety of sizes and shapes, a pitcher, a cookie sheet, and a dry material for pouring.

Show the balance scale, a table-tennis ball, and a golf ball, and explain:

- **This is a scale. Watch what happens when I put a table-tennis ball in one pan and a golf ball in the other.**
- **What happened? The heavy thing pushed the pan down. The light thing went up. Which thing is heavy? Which is light?**

Next, show two containers that have different capacities and a pitcher of something dry to pour.

- **Which container will hold more sand? Let's find out.**

Fill one container with sand, and explain:

- **I filled one container with sand. I am going to pour this sand into the other container.**

Fill the second container from the first container.

- **Which container held more? Which container held less?**
- **Which container held more? Which container held less?**

Continue with both activities with other examples; then let volunteers experiment.

Lesson Follow-Up

If students have difficulty completing this activity, have them complete Activity 41.

Activity 38

Developing **Objective**

Students identify parts and wholes in shapes composed of smaller shapes.

Have available a variety of pattern blocks. You will put them together in different configurations to create a variety of shapes.

Put two right triangles together diagonally to form a rectangle/square. Touch the whole and each part as you tell the students:

- **I made a rectangle/square. The whole rectangle/square has two parts. How many parts are in this rectangle/square?**

Put two triangles together "back to back" to form another triangle. Touch the whole and each part as you tell the students:

- **I made a triangle. The whole triangle has two parts. How many parts are in this triangle?**

Put four squares together 2 by 2. Touch the whole and each part as you tell the students:

- **I made a square. The whole square has four parts. How many parts are in this square?**

Ask a volunteer to put the four squares together to form a tall rectangle. Ask a student to point out the parts and the whole.

Continue combining shapes to create other geometric shapes. Have the students identify the parts and the whole.

Lesson Follow-Up

If students have difficulty completing this activity, have them complete Activity 42.

Conceptual Development Activities

Activity 39

Objective

Students use concepts of time including *morning, afternoon, before, after,* and *next.*

Discuss activities students do at different times during the school day. If you have different morning and afternoon classes, adjust the questions.

- **What do we do in the morning?**

If necessary, lead with more specific questions, such as, "Do we sing a song in the morning? Do you play outside in the morning?"

- **When do we (a morning activity)?**

If necessary, lead with more specific questions, such as, "Do we read books after our snack? Do you go to learning centers before math?"

- **What do we do after (a morning activity)? What do we do next?**
- **What other things do we do in the morning?**

Follow similar lines of questioning for afternoon if you have whole-day classes.

Lesson Follow-Up

If students have difficulty completing this activity, have them complete Activity 43.

Activity 40

Objective

Students describe uses of coins.

Display a variety of and coins, and ask:

- **Have you ever seen any of these things? What are they?**
- **When I want to buy a bottle of juice out of a machine, I put coins into the machine, and it gives me a bottle of juice. I traded money for juice.**
- **Have you had any of these coins? What did you do with them?**
- **When do grown-ups use coins?**

Lesson Follow-Up

If students have difficulty completing this activity, have them complete Activity 44.

Conceptual Development Activities

Activity 41

Developing **Objective**

Students compare objects by weight and describe them as light and heavy.

Have available several objects that are light and others that are heavy; for example, cotton balls, tissue boxes, books, boxes of pebbles or marbles.

Give a student a light object and a heavy object, and ask:

- **Which is more difficult to lift? It is heavy.**
- **Which is easy to lift? It is light.**
- **If something is hard to lift, we say it is heavy. If something is easy to lift, we say it is light.**

Continue giving students pairs of objects, and ask them to decide which is heavy and which is light.

Lesson Follow-Up

If students have difficulty completing this activity, have them complete Activity 45.

Activity 42

Emerging **Objective**

Students recognize parts of objects.

Have available several common classroom objects that have several parts; for example, stuffed animals, dolls, trucks, books, or a computer.

Show the object, and ask the students to identify it, or show the object, and ask a question that can be answered with Yes or No.

Then ask the students to name a part as you touch it. If necessary, touch a part, and ask a question that can be answered with Yes or No.

Then name a part, and ask a student to touch the part.

Make a statement that names the whole and its parts:

- **This is a truck. The whole thing is called a truck. Some parts of the truck are the tires, the cab, the bed, the windows, the doors, and the hood.**

Continue with additional objects.

Lesson Follow-Up

If students have difficulty completing this activity, have them hold the object while you ask them to point to its parts.

Activity 43

Emerging **Objective**

Students describe activities that occur at various times of the day.

Discuss activities students do at different times of day.

- **What do you do in the morning before you come to school?**

If necessary, lead with more specific questions, such as, "Do you brush your teeth in the morning? Do you get dressed in the morning?"

- **When do we (a morning activity)?**

If necessary, lead with more specific questions, such as, "Do we read books in the morning? Do you go to learning centers in the morning?"

- **What do we do at lunchtime?**

Follow similar lines of questioning for afternoon and night.

Lesson Follow-Up

If students have difficulty completing this activity, concentrate on times and time words when you direct students from one activity to another. For example, "This morning we are going to (work with clay). After lunch, we will play outside."

Activity 44

Emerging **Objective**

Students understand the concepts of giving and receiving.

Distribute a variety of small objects, counters, and blocks. Keep some for yourself. Show one of your objects or counters, and initiate a trade:

- **We're going to play a game with these. I will give you something I have, and you give me something you have. (Daniel), let's trade. I'll give you a counter. What will you give me?**

Continue trading with several students or with the same student several times. Then encourage the students to trade with you and one another.

Lesson Follow-Up

If students have difficulty completing this activity, make trading part of daily activities.

Conceptual Development Activities

Activity 45

Emerging **Objective**

Students compare capacities of objects and describe which holds more and which holds less.

Have available containers in a variety of sizes and shapes and a pitcher of water or a dry material for pouring.

Display two containers with very different capacities, such as a small plastic glass and a large empty milk carton. Ask the students:

- **Tell/Show me which container you think will hold more.**
- **Let's find out.**

Fill one container, and explain:

- **I filled one container. I am going to pour this (water) into the other container.**

Fill the second container from the first container. If it holds all the water or material, pour more into it, and ask:

- **Tell/Show me which container held more. Which container held less?**

If the second container did *not* hold all the water or material, stop pouring, and ask:

- **Tell/Show me which container held more. Which container held less?**

Then have a student choose two containers and tell which he/she thinks will hold more. If the student is able to pour, help him/her perform the experiment. If not, do it yourself.

Repeat the activity, and then let the students experiment.

Lesson Follow-Up

If students have difficulty completing this activity, have them complete Comparisons from the ***Building Blocks*** program.

Activity 46

Independent **Objective**

Students model the doubles addition facts through $5 + 5$.

Have available 10 counters, math-link cubes, or blocks for each student and a set for demonstrating.

Tell the students:

- **We are going to play Double It. I will say a number. You make a set of counters for that number. Like this. The number is one.**

Place one counter as a set on the table.

- **Then I will tell you to make a second set like the first set. I will say "Double it."**

Place one more counter as a set on the table, and explain:

- **I made two sets that each have one. That means I doubled one.**
- **Let's join the sets. How much is one plus one?**
- **Now we are going to double two. I will say a number. You make a set of counters for that number. The number is two.**
- **Now double it—make a second set like the first set.**

After students have made a second set of two, continue:

- **How much is two plus two?**

Continue through double fives.

Lesson Follow-Up

If students have difficulty completing this activity, have them complete Activity 47.

Conceptual Development Activities

Activity 47

Developing **Objective**

Students solve problems involving the doubles addition facts 1 + 1 and 2 + 2.

Give each student a group of four counters or blocks. Demonstrate as you say:

- **Here's a story. Lee has a block, and Steve has a block. How many blocks do they have together? Let's make sets to find the answer. Make a set of one block for Lee.**

Provide help if necessary.

- **Make a set of one block for Steve. Now match one block for Lee and one block for Steve.**
- **Join the two sets. How many blocks do they have in all?**

Count the blocks with the students to confirm.

Repeat with two and two.

Lesson Follow-Up
If students have difficulty completing this activity, have them complete Activity 48.

Activity 48

Emerging **Objective**

Students model the doubles addition facts 1 + 1 and 2 + 2.

Have available several small classroom objects, such as counters, blocks, or cars.

Set one object in front of you.

- **Here is one counter. I made a set of one.**

Set a second object in front of you, away from the first counter.

- **Here is one counter. I made another set of one.**
- **Now I am going to join the sets.**

Push the two counters together.

- **I put a set of one with a set of one. Now I have a new set.**
- **Tell/Show me how many in all.**

Give the students two objects, and repeat the procedure with them.

Repeat the procedure with sets of two and two.

Lesson Follow-Up
If students have difficulty completing this activity, have them work with Count and Race from the ***Building Blocks*** program.

NOTES

3

Conceptual Development Activities

Activity 1

Independent **Objective**

Students combine (multiply) equal sets with quantities to 18 using objects and pictures with numerals.

Give students 18 colored erasers. Have students combine the erasers in 6 rows, putting 3 erasers in each row. Next, have students count the number of erasers in each row and the number of rows. Write the number model on the board: $6 \times 3 =$. Ask how many erasers there are in all. Model counting the total number of erasers (18) with the students. Use the erasers to make equal sets with different quantities up to 18. Have students write each number model on the board and count the total number of erasers to solve the problem.

- **How many erasers are in each row?**
- **How many rows of erasers are there?**
- **How many erasers are there in all?**
- **What is the number model?**

Lesson Follow-Up

If students have difficulty completing this activity, have them complete Activities 5 and 6.

Activity 2

Independent **Objective**

Students solve addition facts with sums to 18 and related subtraction one-digit fact families using the formal algorithm with numerals and signs (+, –, =).

Draw a triangle on the board. Write the number of a fact family in each corner of the triangle (for example, 7, 9, and 16). Place the highest number in the top corner of the triangle. Put an addition and subtraction symbol in the middle of the triangle. Explain that these numbers are a fact family. Model on the board $7 + 9 = 16$, $9 + 7 = 16$, $16 - 9 = 7$, $16 - 7 = 9$. Discuss how fact families will help students solve addition and subtraction facts up to 18. Model more fact family triangles and the correlating algorithms on the board.

- **What is a fact family?**
- **Which numbers belong in a fact family?**
- **What is the number model or algorithm for the fact family?**
- **How are the numbers related?**

Lesson Follow-Up

If students have difficulty completing this activity, have them complete Activities 5 and 6.

Activity 3

Independent **Objective**

Students use one-to-one correspondence, grouping, and counting as strategies to solve real-world problems involving addition facts with sums to 18 and related subtraction facts.

Remove the face cards from a standard card deck, or use number cards that show numerals and corresponding picture sets. Place the deck of cards between two students, and have the students draw two cards each. Have them use one-to-one correspondence to count the total number on the cards. The person with the highest number wins the round and keeps both cards. Have students use the count up strategy to find the sum of the two cards. Ask students:

- **Who has the highest number?**
- **Who has the lowest number?**
- **Show me/Tell me how many you have on both cards.**

An alternative way to play is to have the person with the lowest number keep the cards. Have students use the count back strategy to find the difference of the two cards.

Lesson Follow-Up

If students have difficulty completing this activity, have them complete Activity 7.

Activity 4

Independent **Objective**

Students use objects and pictures to represent the inverse relationship between addition and subtraction facts.

Use classroom objects such as blocks, counters, or erasers to represent the inverse relationship between addition and subtraction facts. Model the addition and subtraction facts for students using the blocks.

- **Add 5 blocks and 3 blocks, and you have 8 blocks.**
- **Take 3 blocks away from 8 blocks, and you have 5 blocks.**

As you model the facts, write the number models on the board. Have students use the blocks to represent addition and subtraction facts. Continue to complete the activity with students if needed.

Next, use pictures of shapes, and have students draw more to add or cross out to subtract the shapes. Use number models with each picture problem. Example:

- **Draw 6 circles. Add 3 more circles. Write the addition fact.**
- **Draw 9 circles. Cross out 6 circles. Write the related subtraction fact.**

As you model, ask:

- **How many do you have now?**
- **What is the total amount?**

Lesson Follow-Up

If students have difficulty completing this activity, have them complete Activity 5.

Conceptual Development Activities

Activity 5

Developing **Objective**

Students combine (multiply) equal sets with sums to 9 using objects and pictures.

Give students 9 erasers. Have students combine the erasers in 2 rows, putting 2 erasers in each row. Next, have students count the number of erasers in each row and the number of rows. Write the numbers modeled on the board as $2 \times 2 =$. Ask how many erasers there are in all. Model counting the total number of erasers (4) with the students. Use the erasers to make equal sets with different quantities up to 4. Have students write each number model on the board and count the total number of erasers to solve the problem.

- **How many erasers are in each row?**
- **How many rows of erasers are there?**
- **How many erasers are there in all?**
- **What is the number model?**

Next, use pictures to demonstrate quantities up to 9 with students. Draw 2 groups of 3 coins. Circle groups of three.

- **How many groups are there?**
- **How many coins in all?**
- **What is the number model?**

Lesson Follow-Up

If students have difficulty completing this activity, have them complete Activity 8 or 9.

Activity 6

Developing **Objective**

Students solve addition facts with sums to 9 and related subtraction facts using numerals with objects and pictures.

Draw a triangle on the board. Write the number of a fact family in each corner of the triangle (for example, 1, 4, and 5). Place the highest number in the top corner of the triangle. Put an addition and subtraction symbol in the middle of the triangle. Explain that these numbers are a fact family. Give students objects or use pictures to demonstrate the addition and subtraction facts. Then, model on the board $1 + 4 = 5$, $4 + 1 = 5$, $5 - 1 = 4$, $5 - 4 = 1$. Discuss how fact families will help students solve addition and subtraction facts up to 9. Model more fact family triangles, and use the objects or pictures to solve the problems.

- **What is a fact family?**
- **Which numbers belong in a fact family?**
- **What is the number model or algorithm for the fact family?**
- **How are the numbers related?**

Lesson Follow-Up

If students have difficulty completing this activity, have them complete Activity 8.

Conceptual Development Activities

Activity 7

Developing Objective

Students use one-to-one correspondence, grouping, and counting as strategies to solve real-world problems involving addition facts with sums to 9 and related subtraction facts.

Give students a handful of counters. Tell real-world problems involving addition facts up to 9 and their related subtraction facts. Example:

- **There were 9 students in the classroom. One student left the room. How many students are in the class?**
- **Seven boys are buying lunch. One girl is buying her lunch. How many students are buying lunch altogether?**

Have students use one-to-one correspondence to group and count the total number of counters for each problem. Ask:

- **Show me/Tell me how many counters you had in all.**
- **How many are left?**

Then write the number models on the board. Allow students to choose from number cards to show their responses.

Lesson Follow-Up

If students have difficulty completing this activity, have them complete Activity 8.

Activity 8

Emerging Objective

Students solve simple problems involving joining or separating sets of objects to 3.

Give students three pennies or counters. Model separating and joining the pennies before students do it. Use simple number stories for students to join or separate pennies. Example:

- **Andre had 2 pennies. His mom gave him 1 more penny. Show me/Tell me how many pennies he had in all.**
- **Graham had 3 pennies. He gave 3 pennies to his sister. Show me/Tell me how many pennies he has now.**

After each number story, ask students:

- **Did you have to join or separate your pennies?**
- **How many pennies do you have? Choose the number card that shows your answer.**

Lesson Follow-Up

If students have difficulty completing this activity, use healthy snacks to demonstrate joining and separating. Have students join the snacks and separate the snacks by eating them. (Be aware of food allergies.)

Conceptual Development Activities

Activity 9

Emerging **Objective**

Students recognize when 1 or 2 items have been added to or removed from sets of objects to 3.

Give each student the same amount of counters or coins. Together, count the number of items up to 3. Give each student one more item. Create plus and minus cards for each student. Students will use a + sign to recognize that items have been added or a – sign to indicate if items have been removed from the set of 3. Ask:

- **Did you add or remove items?**
- **Show me/Tell me how many you have now.**

Lesson Follow-Up

If students have difficulty completing this activity, have them use other types of objects that are important to students (favorite toys, blocks, crayons). Continue doing the activity with the students until the students understand.

Activity 10

Independent **Objective**

Students represent half and whole using area and sets of objects.

Place 6 pencils on the table. Count the number of pencils with the students. Have a student take away three of the pencils. Ask:

- **How many pencils were taken away?**
- **How many are left on the table?**

Discuss with students how 6 pencils were the whole set of pencils and 3 of the pencils represented half of the objects. Use more examples with different classroom objects such as books, crayons, and markers. Ask:

- **Show me half of the objects.**
- **Show me all (whole) of the objects.**

Next, fold a piece of paper in half, and color one half of the paper. Ask:

- **How much of the paper is colored?**

Discuss with students how the two halves of the paper make up the whole.

Lesson Follow-Up

If students have difficulty completing this activity, have them complete Activity 12.

Conceptual Development Activities

Activity 11

Independent **Objective**

Students identify the relationship between half and whole.

Give students a variety of paper shapes (square, rectangle, circle, rhombus). Fold the shape in half, and point to each half of the shape. Then have students unfold their shape to identify the whole shape. Continue folding and unfolding the shapes and identifying the relationship between half and whole.

- **Who can show me half of a circle?**
- **Who can show me the whole circle?**

Lesson Follow-Up

If students have difficulty completing this activity, have them complete Activity 12.

Activity 12

Developing **Objective**

Students recognize part and whole using area and sets of objects.

Give students 5 pennies or counters. Have them count the total number of pennies. Then have students separate 4 pennies. Ask what number of pennies was separated from the group. Discuss how four of the pennies are part of the whole set of 5 pennies. Continue this activity, with students using different amounts of pennies. Discuss the parts and whole sets of objects with students.

- **How many pennies are there in all?**
- **What part of the pennies was separated from the whole group?**

Next, fold a piece of paper in half and then in half again to create four equal parts. Discuss with students how the four parts of the paper make up the whole. Then color one part. Ask:

- **How many parts in all?**
- **What part is colored?**

Repeat, coloring different amounts of paper.

Lesson Follow-Up

If students have difficulty completing this activity, have them complete Activity 13.

Conceptual Development Activities

Activity 13

Emerging **Objective**

Students recognize parts of whole objects and parts of sets of objects.

Give students a square with lines to show 3 equal parts of the whole square. Have students color one part of the square, assisting them as necessary. Ask:

- **Show me/Tell me which part is colored.**

Tell students that one out of three parts is colored. Use a variety of different paper shapes divided into equal parts. Have students color the parts and show/tell you what they did. Use number cards, if necessary. Ask:

- **How many parts of the shape are colored?**

Repeat the activity with sets of classroom objects.

Lesson Follow-Up

If students have difficulty completing this activity, have them focus on representing half of the whole object. Color the halves of the square, and outline the entire square with a black marker.

Activity 14

Independent **Objective**

Students identify attributes, including number of sides, curved or straight sides and number of corners (angles) in two-dimensional shapes.

Use a 16-box bingo board, and draw a variety of two-dimensional shapes with curved or straight sides and various numbers of sides in each box. Copy your board for all students. Give bingo boards and counters to students. Tell students an attribute of one of the shapes. For example, say, "Put a counter on the shape with curved sides." Students would put the counter on the box with the circle in it. The winner is the person who gets four in a row first and calls out "Bingo."

Discuss the attributes with students by asking:

- **What shape has curved sides?**
- **What shape has 4 sides?**
- **What shape has 3 angles?**

Lesson Follow-Up

If students have trouble with this activity, have them complete Activity 17.

Conceptual Development Activities

Activity 15

Independent

Objective

Students combine (compose) and separate (decompose) two-dimensional shapes to make other shapes.

Draw pictures of two-dimensional shapes on the board, and label each one. Give students pattern blocks. Have students use the pattern blocks to compose various shapes, such as a rectangle. Then have the students decompose the shapes to make other shapes, such as two squares or two triangles. Ask questions such as the following:

- **How many sides does a triangle have?**
- **How many squares/triangles will you need to make a rectangle?**
- **Tell me/Show me how you compose a rectangle out of pattern blocks.**

Lesson Follow-Up

If students have difficulty completing this activity, have them complete Activity 18.

Activity 16

Independent

Objective

Students identify two-dimensional shapes that are the same shape and size (congruent).

Explain to the class the definition of *congruent*. *Congruent* means shapes are the same size and shape. Give students a variety of attribute blocks, making sure that there is at least one pair of congruent shapes in the blocks. Have students manipulate the blocks by placing them on, next to, and on top of each other to see if they are congruent.

- **How do we know if a shape is congruent?**

Lesson Follow-Up

If students have trouble with this activity, have them complete Activity 19.

Conceptual Development Activities

Activity 17

Developing **Objective**

Students sort two-dimensional shapes by single attributes, including numbers of sides and curved or straight sides.

Give students paper shapes of circles, triangles, rhombuses, pentagons, rectangles, and squares. Have students sort the shapes by single attributes, including the number of sides or curved versus straight sides. Ask:

- **What shape had a curved side?**
- **What shape had more than one side?**
- **What was the shape with the most sides?**

If students have trouble remembering the attributes they are sorting by, consider giving them labeled mats to work on.

Lesson Follow-Up
If students have trouble with this activity, have them complete Activity 20.

Activity 18

Developing **Objective**

Students combine (compose) two dimensional shapes to make other shapes.

Draw pictures of two-dimensional shapes on the board, and label each one. Give students straws. Have students compose two-dimensional shapes with the straws, and have them glue the shapes on construction paper and label each one with the number of sides. Then have students create a second shape to add to the first one, and discuss their results.

- **What shapes can you make with the straws?**
- **How many sides does your shape have?**
- **How many angles?**
- **Are the sides straight or curved?**

If students have trouble working with the straws, have them use pattern blocks to compose shapes.

Lesson Follow-Up
If students have trouble with this activity, have them complete Activity 20.

Conceptual Development Activities

Activity 19

Developing **Objective**

Students match two-dimensional shapes that are the same shape and size (congruent).

Explain to the class the definition of *congruent*. *Congruent* means shapes are the same size and shape. Give students various shapes to cut out, or have shape cutouts already prepared. Have students match the shapes that are congruent.

- **Show me/Tell me which shapes are congruent**
- **Show me/Tell me which shapes are not congruent.**
- **How do you know?**

Alternatively, have students use real-world objects such as a coffee-can lid or a fast-food drink lid. Compare the lids to see if they are congruent. They are the same shape, but they are not the same size. Continue using real-world objects to find shapes that are or are not congruent.

Lesson Follow-Up

If students have difficulty with this activity, have them complete Activity 21.

Activity 20

Emerging **Objective**

Students recognize common objects with two-dimensional shapes, such as circle and square.

Tell students to look for objects around the classroom that have the shape of a circle or a square. Have students go on a shape hunt around the room. Students could point to a clock (circle), desk (square), coins (circle), and so on. Make a chart of all the common objects that students found and their shapes.

- **Who can find something that is a circle?**
- **Show me/Tell me an object that is in the shape of a rectangle or square.**

If students are unable to move around the classroom, have common objects at the table, and have the student choose the circle from the objects presented.

Lesson Follow-Up

If students have difficulty completing this activity, identify the common objects (such as books, sticky notes, or money), and ask students the shapes of the objects.

Conceptual Development Activities

Activity 21

Emerging **Objective**

Students recognize two-dimensional shapes, including circle and square, that are the same shape and size (congruent).

Take a walk around the school or classroom to look for different shapes. Record the shapes on paper. Help students compare objects with similar shapes to see if they are the same size. Point out to students some of the following objects: books, CDs, DVDs, videos, notebooks, calendars. Ask:

- **Who can find a square?**
- **Tell me/show me where to find a square that is the same size.**
- **What is the same shape and size as this circle?**

Lesson Follow-Up

If students have difficulty completing this activity, identify the common objects (such as books, sticky notes, or money), and ask students the shapes of the objects before having them compare the sizes of the objects.

Activity 22

Independent **Objective**

Students complete growing visual and number patterns.

Give students a sticky note. Draw a visual pattern on the board using shapes (circle, square, rectangle, square). Have students complete the visual pattern by drawing the shape that goes next. Next, write a number pattern on the board (count by 2s, 5s, or 10s). Have students use the sticky note to complete the number pattern. Give students an opportunity to create a number or visual pattern independently. Pair-Share: Have students see if a partner can complete the growing number pattern.

- **Show me/Tell me what shape comes next.**
- **Show me/Tell me what number comes next.**
- **How do you know?**
- **What is the pattern of the shapes (numbers)?**

Lesson Follow-Up

If students have trouble with this activity, have them complete Activity 23.

Conceptual Development Activities

Activity 23

Developing **Objective**

Students match a two-element repeating pattern using objects and pictures.

Explain to the class that a pattern repeats objects over and over. Display a red and yellow pattern using counters on the table. Give students red and yellow counters, and have them match the pattern on the table. Use other objects around the room such as erasers, crayons, cubes, or coins to make more two-element repeating patterns.

- **What two-element repeating pattern did we make?**
- **Show me/Tell me what color/object/picture would come next.**
- **How do you know?**

Lesson Follow-Up

If students have trouble with this activity, have them complete Activity 24.

Activity 24

Emerging **Objective**

Students recognize the next step in a simple pattern or sequence e of activities.

Explain to the class that a pattern repeats itself again and again. Work with students to make a clap pattern:

- **Every time we switch activities, we are going to clap a pattern.**

Use a clapping pattern before the students transition from one activity to the next. Clap once; pause; clap twice. Wait for students to respond with the clapping pattern.

- **What is the clapping pattern?**
- **Show me/Tell me what comes next.**

Lesson Follow-Up

If students have difficulty with this activity, give them a visual model of a washing hands routine. Give students picture cards to put the repeating pattern in order. Students will match the simple pattern by placing the matching cards underneath the correct visual model cards.

Conceptual Development Activities

Activity 25

Independent **Objective**

Students use a ruler to solve problems involving the length of sides of squares and rectangles.

Explain to students how to use a ruler, and explain how the ruler can measure in centimeters and inches. Give students various sizes and colors of squares and rectangles. Have students estimate the lengths of the sides in inches. Write the estimate. Next, have students line up the ruler correctly to measure the length of each side. Have them solve problems.

- **Which square or rectangle is the largest?**
- **What is the total of all the sides?**
- **What is the total of two sides?**

Lesson Follow-Up

If students have trouble with this activity, have them complete Activity 29.

Activity 26

Independent **Objective**

Students identify half and whole of the length of objects.

Collect several classroom objects such as pencils, books, and crayons. Use a ruler to measure the length of the object. Write the total amount. Identify the whole and half length of the object. For example, measure the length of a new pencil. It is 6 inches long. Help students figure half the length of the object. Record both whole and half lengths.

- **What is the half length of the object?**
- **What is the whole length of the object?**

Lesson Follow-Up

If students have difficulty completing this activity, have them complete Activity 30.

Conceptual Development Activities

Activity 27

Independent **Objective**

Students identify time to the hour and half hour using analog and digital clocks.

Make a bingo board with 16 boxes (4 boxes on each side). Write digital times to the half hour and hour. Use a demonstration clock to display analog times to students in the class. Have students identify the time to the half hour and hour, and use a bingo marker to match the digital and analog times. The first person who gets 4 in a row wins. Ask:

- **Where is the hour hand?**
- **Where is the minute hand?**
- **What time is it? How do you know?**

Lesson Follow-Up

If students have difficulty with this activity, have them complete Activity 31.

Activity 28

Independent **Objective**

Students identify the months of the year in relation to calendars.

At the beginning of the year, set up a calendar routine. Use the names of the months with picture clues to help students identify each month. (For example, flowers represent May; a four-leaf clover represents March.) Make a calendar page for each month. Number the days. Have students label the month at the top of the page and name any special events that month.

- **What is this month?**
- **What month was last month?**
- **What month is next month?**

Lesson Follow-Up

If students have difficulty with this activity, have them complete Activity 32.

Conceptual Development Activities

Activity 29

Developing **Objective**

Students use nonstandard measurement units to solve problems for length of sides of squares.

Tell students they are going to use large paper clips to measure the lengths of squares. Demonstrate how to lay the paper clips end-to-end to measure each side. Have students compare which squares used the most paper clips. Discuss how many paper clips students used per side. Discuss how many paper clips they used in all. Ask:

- **How many paper clips did you use to measure one side of the square?**
- **How many paper clips did you use to measure all of the sides?**

Next, have students use a different-sized object, such as a smaller paper clip or a crayon, to measure the same squares. Discuss the results. Ask:

- **How many small paper clips were used to measure one side?**
- **How many large paper clips were used to measure the same side?**

Lesson Follow-Up

If students have trouble with this activity, have them complete Activity 33.

Activity 30

Developing **Objective**

Students recognize part and whole of the length of objects.

Collect several classroom objects such as pencils, books, and crayons. Use a ruler to measure the length of an object. Write the total amount. Then identify the whole length and the different parts of the length of the object. For example, measure the length of a new pencil. It is 6 inches long. Help students understand that each inch is one part of the length of the object.

- **What is part of the length of the object?**
- **What is the whole length of the object?**

Lesson Follow-Up

If students have trouble with this activity, have them complete Activity 33.

Conceptual Development Activities

Activity 31

Developing Objective

Students identify concepts of time, including yesterday, today and tomorrow, by relating activities to the time period.

Sing a song with students every day. "Today is (day). Yesterday was (day). What will tomorrow be?" Have the song written on poster board. Use days-of-the-week word cards to fill in the blanks on the song. Relate activities to the time period with the students. For example, say, "Yesterday we went on a field trip. Today we are going to gym. Tomorrow is Saturday, so you won't come to school." Ask:

- **What day is today?**
- **What are we going to do today?**
- **What did we do yesterday?**

Lesson Follow-Up

If students have difficulty completing this activity, have them complete Activity 34.

Activity 32

Developing Objective

Students identify the days of the week using a calendar.

At the beginning of the year, set up a calendar routine. Sing the days of the week each day. Have students practice identifying the current day by pointing to the right day. As the year progresses, give students days-of-the-week word cards, and have them practice putting the days of the week in order on the calendar each day. Write the month, day, and year on the board each day. Ask:

- **What day is today?**
- **What day will tomorrow be?**
- **How many days are in a week?**
- **What is the first day of the week?**

Lesson Follow-Up

If students have difficulty completing this activity, have them complete Activity 34.

Activity 33

Emerging Objective

Students recognize the sides of a square or rectangle.

Draw various sizes of squares and rectangles on the board. Draw each side of the shapes with a different color. Explain to your students that a square or a rectangle is made up of 4 sides. Show that there is a different color for each side of the shape. Have students take out four crayons. Have students recognize the sides of a square or rectangle by drawing each side of the shape with a different color or by pointing to each side. If students are unable to draw, have them use different colored strips of paper to create a square or rectangle.

- **How many sides of a square?**
- **Who can point to the sides of a square or rectangle?**

Lesson Follow-Up

If students have difficulty with this activity, have them circle the sides of a square or a rectangle you drew. Prompt students as needed.

Activity 34

Emerging Objective

Students recognize part of day, such as morning or afternoon, associated with a common activity.

Review with the class that morning and afternoon are parts of the day. Give students picture cards with common activities that students complete in the morning and afternoon (eat breakfast, get dressed, brush teeth, come to school, eat lunch, have recess, learn reading, learn math). Have students sort the picture cards into two categories—morning and afternoon activities.

- **What activities do we do in the morning?**
- **What activities do we do in the afternoon?**
- **What are two parts of the day?**

Lesson Follow-Up

If students have difficulty completing this activity, have students work with pictures of only two activities to sort. Give prompts so students sort the picture cards accurately.

Conceptual Development Activities

Activity 35

Independent Objective

Students express, represent and solve problems with cardinal numbers 0 to 30 and ordinal numbers to tenth using sets of objects or pictures, number names and numerals.

Explain that ordinal numbers show the order of things (first, second), and cardinal numbers tell the amount of something (one, two, three). Make a set of cardinal and number name cards to 30 and a set of ordinal number cards to tenth. For example, make a card with the number 1 on it and the word *one*. Then make a separate card with the word *first*. Have students show the matching cardinal and ordinal numbers up to ten using the numerals and number name cards.

Next, give students simple problems to solve involving common activities. Allow students to use counters to help them solve the problems. For example, say:

- **Four players are playing a game. What is the order of each player? (first, second, third, fourth).**
- **I have 15 books. I loan four books to Teresa. How many books do I have left?**

Each time have students use their ordinal numerals and number name cards to express the answer to the problem when appropriate. Continue having students solve problems with cardinal numbers 0–30 and ordinal numbers *first* through *tenth*.

Lesson Follow-Up

If students have difficulty with this activity, have them complete Activity 37.

Activity 36

Independent Objective

Students apply the concepts of counting and grouping to create sets of tens and ones to identify the value of whole numbers to 30.

Give students a set of 30 craft sticks and a ones and tens chart. Have students put all the craft sticks in the ones column. Explain to students they have 30 craft sticks. Write 30 on the board. Have students count out 10 of the sticks and put a rubber band around them. Explain to students there is 1 group of 10 craft sticks. Put this group in the tens column. Have students count all the craft sticks in the ones and tens columns. Write the amount on the board (30). Have students count 10 more craft sticks that are in the ones column and put a rubber band on them. Say, "We have 2 groups of 10 craft sticks. What does that equal together?" (20) Have students count the total number of craft sticks in the ones and tens columns. Write the amount on the board. Have students count 10 more craft sticks that are in the ones column and put a rubber band on them. Say, "We have 3 groups of 10 craft sticks." Write the total amount on the board. (30) Throughout the activity, ask:

- **How many craft sticks are in the ones?**
- **How many craft sticks are in the tens?**
- **Who can count by tens?**

Lesson Follow-Up

If students have trouble with this activity, have them complete Activity 38.

Conceptual Development Activities

Activity 37

Developing **Objective**

Students express, represent and solve problems with numbers to 10 using sets of objects, pictures, number names and numerals.

Give students a set of 10 miniature race cars and a set of number cards that contain the numeral, number name, and corresponding picture on each card (for example, the number 10, the word *ten*, and ten dots). Give students simple problems to solve with the race cars.

- **I have 5 race cars. I buy 2 more. How many in all? Show me the number card that matches that amount.**

Continue the activity with additional problems involving numbers to 10. Have students use the cars to solve the problems and the cards to express and represent the answers. Ask:

- **How many cars are there?**
- **What number card matches the amount of cars?**

Lesson Follow-Up

If students have difficulty completing this activity, have them complete Activity 39.

Activity 38

Developing **Objective**

Students use one to one correspondence to count sets of objects to 10.

Give students a set of 10 craft sticks. Have students put the sticks in groups of ten, check their groups by matching the number of craft sticks to a dot card or number card, and put a rubber band around each group. Repeat with other quantities up to 10. Ask:

- **How many are in the rubber band?**
- **Who can put the craft sticks into a group of 10?**

Lesson Follow-Up

If students have difficulty completing this activity, have them complete Activity 40.

Conceptual Development Activities

Activity 39

Emerging Objective

Students recognize quantities 1 to 3 using sets of objects, pictures or number names.

Give students three containers marked with a number name, a picture of the amount, and the numeral (for example, *three, 3,* and three dots). Have students use counters, small teddy bears, or erasers to show the quantity up to 3.

- **Show me three teddy bears, and put them in the matching container (or point to the container that says 3).**
- **Show me two teddy bears, and put the teddy bears in the matching containers.**
- **Which container has three teddy bears?**

Lesson Follow-Up

If students have trouble with this activity, have the students use the teddy bears to count up to 3. Don't use the matching containers. Model on the board the amount the students counted with just the number and a picture of the number.

Activity 40

Emerging Objective

Students match objects to marked spaces to show one-to-one correspondence e for quantities to 3.

Model counting with one-to-one correspondence up to 3 using a three-frame card and counters. Give students a three-frame card and three counters. Have students count with one-to-one correspondence by placing the counters on the three-frame card. Give students different amounts to show (1, 2, 3), and have them place that number of counters on the three-frame card. If students are unable to place the counters on the card, place them for them, and give them a choice of number cards to match to the set. Ask:

- **Show me/Tell me how many counters are on the card.**

Lesson Follow-Up

If students have trouble with this activity, have them use two gloves and match each one to their hands. Count with one-to-one correspondence as they put on each glove.

Conceptual Development Activities

Activity 41

Independent **Objective**

Students sort and count objects and pictures into three labeled categories and display data in an object graph or pictograph.

Give students a handful of dimes, nickels, and pennies. Have students sort the coins into three groups. Use a piece of graph paper for students to display the coins in the three labeled categories. Next discuss the data on the three-category object graph. Ask:

- **How many dimes are there?**
- **How many pennies are there?**
- **How many nickels are there?**
- **Do you have more pennies or dimes?**
- **Compare your object graph with a graph done by another student.**

Lesson Follow-Up

If students have difficulty completing this activity, have them complete Activity 42.

Activity 42

Developing **Objective**

Students sort objects representing data into two labeled categories and count the number in each category.

Give students five yellow counters and four red counters. Have students sort the counters by color, count the number of counters in each category, and then label each category.

Alternatively, have students vote on their favorite color crayon (red or blue), by placing a red or blue crayon in a container to represent their vote. Then have students sort the crayons by color and count the data. Ask:

- **How many objects were in each group?**
- **Which group had the most objects?**

Lesson Follow-Up

If students have difficulty completing this activity, have them complete Activity 43.

Activity 43

Emerging **Objective**

Students identify items that belong together to form a set (data).

Take a vote with students about their favorite color. Use colored construction paper to indicate the choices. Students may use a picture of themselves to vote for their favorite color by placing it under the corresponding color. Then help students count the number of pictures that go together (data).

- **Which color had the most votes?**
- **Which color had the least votes?**

Lesson Follow-Up

If students have difficulty completing this activity, have them use everyday objects such as pencil and paper, pairs of socks, or toothbrush and toothpaste. Have them identify the items that belong together.

NOTES

Conceptual Development Activities

Activity 1

Independent Objective

Students informally multiply and divide by combining and separating sets to 30.

Explain that combining/multiplying is like fast addition. Have students draw 2 triangles and count the angles or corners of each triangle. On a chart, record the number of triangles (2), the number of corners each triangle has (3), and the total number of corners (6). Next, write the corresponding multiplication fact on the board: $2 \times 3 = 6$. Repeat the activity with different shapes for quantities up to 30.

Explain that separating/dividing is the opposite of combining (multiplying) and is like fast subtraction. Give students simple problems to solve that involve separating up to 30 counters into equal sets. For example:

- **Marcy has 6 bones. Marcy has 3 dogs. How many bones will each dog get if Marcy separates the bones equally?**

Write the corresponding division fact on the board: $6 \div 3 = 2$. Then repeat the activity with different quantities up to 30.

- **What is the total amount you started with?**
- **How many times do you need to divide it?**
- **How many did each dog get?**

Lesson Follow-Up
If students have difficulty with this activity, have them complete Activity 3.

Activity 2

Independent Objective

Students add and subtract numbers to 30 to solve real world problems and check for accuracy.

Play Math Baseball. Before beginning the game, draw a baseball diamond on the board. Assign each student a marker to move around the bases. Model adding or subtracting two-digit numbers to 30. Show students how to start adding in the ones column. Add the numbers to find the sum. Then add the numbers in the tens column to find the sum. If students get the answer correct, they move to first base. The first person to make it around all the bases wins. Show students how to use a calculator to check their answers for accuracy.

To play, provide real-world two-digit addition and subtraction problems for students to solve; for example, "There are 14 chairs in the lunchroom. If we add 11 more chairs, how many chairs will there be altogether?" As students solve their problems, ask:

- **What is the total amount?**
- **What is the math fact?**
- **What happens to the amount when you add or subtract?**

Lesson Follow-Up
If students have difficulty completing this activity, have them complete Activity 4.

Conceptual Development Activities

Activity 3

Developing **Objective**

Students model multiplication and division facts using up to 15 objects.

Give each student 6 paper clips. Have students arrange the paper clips in 3 rows of 2 paper clips. Next, have students count the number of paper clips in each row and the number of rows. Write the number model on the board: $3 \times 2 = ?$ Ask how many paper clips there are in total. Model counting the total number of paper clips with the students. Use the paper clips to make equal sets with different quantities up to 15. Have students write the number model on the board. Count the total number of paper clips to solve the problem.

- **How many paper clips are in each row?**
- **How many rows of paper clips are there?**
- **How many paper clips are there in all?**
- **What is the number model?**

After students have worked several multiplication problems, model division. Start with a group of 6 paper clips, and ask students to divide the paper clips into 3 equal sets or rows. (If students have trouble, suggest they use one-to-one correspondence to check whether their sets are equal.) When students are ready, write the number model on the board: $6 \div 3 = 2$.

- **How many paper clips did you start with?**
- **How many rows do you have?**
- **How many paper clips are in each row?**
- **What is the number model?**

Repeat the activity with different quantities up to 15.

Lesson Follow-Up

If students have difficulty completing this activity, have them complete Activity 5.

Activity 4

Developing **Objective**

Students create and solve addition and subtraction equations for real world problems involving numbers up to 15.

Give students number cards from 0 to 15 that show both a numeral and a corresponding set picture (for example, the numeral 9 with 9 dots). Also provide cards with the +, −, and = signs. Then give students real-world addition and subtraction problems to solve involving quantities up to 15. For example:

- **I brought 8 oranges to the class picnic. Tammi brought 4 oranges. How many oranges did we have altogether?**
- **I had 12 dollars before I went shopping. Now I have 5 dollars. How much money did I spend shopping?**

Have students use the number and symbol cards to create equations for each problem. As students solve the problems, ask:

- **What do you know about the problem?**
- **What do you want to find out?**
- **Will you add or subtract?**
- **What is the math fact?**

Lesson Follow-Up

If students have difficulty with this activity, have them complete Activity 6.

Conceptual Development Activities

Activity 5

Emerging **Objective**

Students separate and join objects to 4 to solve problems.

Distribute 4 counters or classroom objects to each student. Model separating and joining the objects. Then use simple number stories for students to join or separate the objects. For example:

- **Ruthie had 2 counters. Her mom gave her 2 more counters. Show me/Tell me how many counters she has in all.**
- **Freddy had 4 books. He gave 3 books to his sister. Show me/Tell me how many books he has now.**
- **Did you have to join or separate your objects?**
- **How many objects do you have?**

If students are unable to manipulate the objects, provide picture cards of each of the objects.

Lesson Follow-Up

If students have difficulty completing this activity, have them use healthful snacks to demonstrate joining and separating.

Activity 6

Emerging **Objective**

Students associate the plus and minus symbols with adding to and removing from a set of objects to 4.

Collect items for students to manipulate (small toy animals, small teddy bears, counters, coins, and so on) and a visual cue card of the + and – signs. Give students a set of 4 objects, and have students count the objects. Ask students to remove one of the objects from the set. Have students use the + or – visual cue card to describe what they did. Give students more practice in adding or removing items from their set and pairing it with the + or – visual cue card.

- **Did you add something?**
- **Did you take something away?**
- **Show me/Tell me what you did.**

Lesson Follow-Up

If students have difficulty with this activity, use paper flowers and stems. Have students place the flower on each stem with Velcro and then remove the flowers from the stems.

Conceptual Development Activities

Activity 7

Independent **Objective**

Students group objects to 50 into sets of ten and ones as a counting strategy.

Use unifix cubes or base-ten blocks and a ones and tens chart. Have students build a set of ten cubes by connecting blocks. Ask students how many individual cubes they used to make the set of ten cubes. Have students continue building sets of 10 using the unifix cubes, put them in the tens column of the chart, and put several individual blocks in the ones column. Have students count the total number in the tens and ones columns and record the whole numbers. For example, "I have 39 because I have three sets of 10 and 9 individual blocks, which equals 39." Ask students to create sets of tens and ones for various whole numbers to 50. As they work, ask:

- **How many sets of 10?**
- **How many ones?**
- **How many altogether?**

Lesson Follow-Up

If students have difficulty completing this activity, have them complete Activity 10.

Activity 8

Independent **Objective**

Students represent halves, fourths, and wholes using sets, area, fractional names, and models.

Use fraction circles to express and represent fractions including halves and fourths. Give students paper circles to fold into halves. Draw a line with the ruler to show the fold creases. Next label each part of the circle.

- **What is the fraction used to show 1 out of 2? To show 2 out of 2?**

Next, have students fold their circle in half again to represent fourths. Draw a line with the ruler to show the fold creases. Label each part of the circle.

- **What is the fractional name for 2 out of 4?**
- **What is the fractional name for 4 out of 4?**
- **What is the fraction for 1 out of 4?**

Next, give students a set of two objects. Demonstrate that there are two equal parts to the set and that each part is $\frac{1}{2}$ of the set. (You may want to place the objects on the fraction circle students made earlier to help them make the connection.) Repeat with a set of 4.

- **How many parts are in each set?**
- **What is each part called?**

Lesson Follow-Up

If students have difficulty completing this activity, have them complete Activity 11.

Conceptual Development Activities

Activity 9

Independent **Objective**

Students represent halves, fourths, and wholes using models.

Give students three square paper shapes. One shape should be left whole, one shape will be folded in half, and the last shape will be folded into fourths. Have students fold one square into fourths. Color each fourth a different color. Have students fold a second square in half. Color each half a different color. Have students color in the last square entirely. Have students identify the shapes by halves, fourths and a whole. Write the fractional name on each square.

Have students compare fractions.

- **Which shape is shaded in fourths?**
- **Which shape is shaded in half?**

Lesson Follow-Up

If students have difficulty completing this activity, have them complete Activity 12.

Activity 10

Developing **Objective**

Students use pennies and dimes to group objects into sets of ten as a counting strategy.

Review with students that a dime is worth 10 cents and a penny is worth 1 cent. Organize students into pairs, and distribute money kits with dimes, pennies, and a number cube. Have students roll the die. The number they roll equals the number of pennies they may take from their money kit. Have students count the number of pennies they have. Once they have 10 pennies, students can trade them in for a dime. Students will continue rolling and trading pennies for dimes until they get to 18 cents. Have students practice counting dimes and pennies.

- **How many dimes do you have?**
- **How much is a dime worth?**
- **How many pennies do you have?**
- **What is the total amount of all the dimes and pennies?**

Lesson Follow-Up

If students have difficulty completing this activity, have them complete Activity 13.

Conceptual Development Activities

Activity 11

Developing **Objective**

Students represent half and whole using sets, area, fractional names, and models.

Use dry-erase boards and markers with students. Have students draw a set of 4 circles. Have students color two of the four circles. Write the fractional name to represent the number of circles they colored in. Continue the activity using varying numbers of circles, such as 3 out of 6. Each time, have the students represent half of the whole set of circles and write the fractional name next to it.

- **What is the fraction for 2 out of 4?**
- **What is the fraction for 4 out of 4?**

Repeat the activity by having students draw one large circle and divide it into two equal parts. Have students color $\frac{1}{2}$ of the circle. Write the fractional name next to it.

Lesson Follow-Up

If students have difficulty completing this activity, have them complete Activity 14.

Activity 12

Developing **Objective**

Students match halves of shapes to create whole shapes and label the halves and wholes.

Make Fraction Puzzles. Give students several paper rectangles, squares, and circles that are already cut in half. Have students match the halves of each shape and then glue the halves onto paper to create the whole shape. Then have students label each part of the shape.

- **Show me half of the rectangle puzzle.**
- **Point to the whole rectangle puzzle.**

Lesson Follow-Up

If students have difficulty completing this activity, have them complete Activity 15.

Conceptual Development Activities

Activity 13

Emerging Objective

Students use one-to-one correspondence to count objects using quantities up to 4.

Make a four-frame card for each student with a large dot on each square. Use coins to have students count with one-to-one correspondence for various quantities up to 4. Have students place the coins on the dots as they count.

- **How many squares are filled?**
- **Show me/Tell me how many coins you have.**

Lesson Follow-Up

If students have difficulty with this activity, have them use common everyday objects (foot and shoe or hat and head) to match objects one-to-one up to 2.

Activity 14

Emerging Objective

Students match part and whole sets of objects to pictures of whole sets and pictures of parts of a set.

Play Part and Whole Match. Give students pictures of several parts and whole sets of objects (for example, pictures of a piece of cake and a whole cake; a piece of pie, whole pie; pizza, slice of pizza). Students should recognize which pairs of objects belong together and glue them to a piece of paper. Discuss with students which pairs of objects belong together.

- **What is this?**
- **What does part of this look like?**
- **Show me/Tell me the picture of the part and the picture of the whole.**

Lesson Follow-Up

If students have difficulty with this activity, have students color parts of a shape picture. Support the student by pointing out the parts of the shapes by the ones they colored in.

Conceptual Development Activities

Activity 15

Emerging **Objective**

Students use objects to identify the whole set and half of the set.

Give students 2 unifix cubes or connecting blocks. Have students count the total number of cubes. Then have students separate the cubes into 2 groups. Discuss how 1 cube is half of the whole set of 2 cubes. (Students may match the cubes one to one if necessary to help them understand that the two sets are equal.) Continue this activity with students using different amounts of cubes. Discuss the halves and whole sets of objects with students.

- **Show me/Tell me how many cubes are in the whole set of cubes.**
- **Show me/Tell me how many cubes make half the set.**

Lesson Follow-Up

If students have difficulty completing this activity, have them use only 2 cubes or connecting blocks. You could also do the activity with pairs of familiar objects, such as mittens or shoes.

Activity 16

Independent **Objective**

Students measure perimeter of a rectangular object using nonstandard units and measure area using square units.

Explain that the distance around all sides of a shape or object is called the perimeter. Use footsteps to measure the perimeter of the school, a hallway, a classroom, or a welcome mat. Record on the chart the perimeter of each item. Have students identify other examples of the distance around all the sides of objects in the environment such as the distance around a basketball court.

Next, explain that area is the space inside a two-dimensional shape. Use one-inch squares to cover a classroom object, such as a book. Explain that these squares show the area of the book cover. Have students identify other examples of area in the environment.

- **What is the perimeter of the shape?**
- **What is the area of the shape?**

Lesson Follow-Up

If students have difficulty completing this activity, have them complete Activity 19.

Conceptual Development Activities

Activity 17

Independent Objective

Students measure the length of the sides of a rectangular object and find the area by counting the square units.

Explain that area is the space inside two-dimensional shapes. Have students measure the length and width of a large rectangular area with carpet squares or construction paper cut into 12-inch squares. Have students count the number of square units inside the rectangular area and the number of squares used on each side of the rectangular area. Record the results.

- **How many squares filled the entire area of the rectangular area?**
- **How many squares were used for the length of each side?**

Lesson Follow-Up

If students have difficulty completing this activity, have them complete Activity 20.

Activity 18

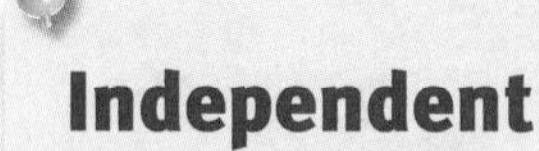

Independent Objective

Students use whole inches and whole feet to measure the length of real life objects.

Explain to students that a ruler uses one-inch markings, and that a ruler is 12 inches, or 1 foot, long. Then have students use a ruler to solve real-world linear measurement problems. Have students measure objects in the classroom like a desk, chalkboard, pencil, crayon, calculator, supply box, book or folder. Record the measurements in whole inches.

- **How many inches long is a book?**
- **How could you measure something that is longer than the ruler?**
- **How many feet long is the chalkboard?**

Lesson Follow-Up

If students have difficulty completing this activity, have them complete Activity 21.

Conceptual Development Activities

Activity 19

Developing **Objective**

Students find real life examples of the concept of perimeter and area.

Explain that area is the space inside a two-dimensional shape. Hold up a piece of paper. Run your finger around the outside of the paper to identify the edge, and then explain that everything inside the edges of the paper is the area. Have students identify other examples of area in the classroom or outside (such as a desktop or tabletop, a floor tile, the playground surface, and so on).

- **Where are the edges of the shape?**
- **Where is the area of the shape?**

Show students various pictures with the perimeter outlined or the area shaded.

- **Which pictures show perimeter?**
- **Which pictures show area?**

Lesson Follow-Up

If students have difficulty completing this activity, have them complete Activity 22.

Activity 20

Developing **Objective**

Students informally calculate the area of a rectangle by counting the number of square units using a grid.

Explain that area is the space inside two-dimensional shapes. Use grid paper that can accommodate shapes up to 10 squares by 10 squares. Have students use rectangle templates of different sizes and trace them on the grid paper or use a ruler to draw various sizes of rectangles. Provide pre-cut rectangles for students, if needed. Have students count the number of squares in each rectangle to determine its area.

- **What is the area of the rectangle?**
- **What did you do to find the area?**

Lesson Follow-Up

If students have difficulty completing this activity, have them complete Activity 22.

Conceptual Development Activities

Activity 21

Developing **Objective**

Students use whole inches to measure the length of the sides of rectangles.

Explain that a ruler uses one-inch markings and that a ruler is 12 inches long. Modify a ruler by using a permanent marker to darken the inch marks. Color the starting place in green. Draw several rectangles, and have students measure the lengths of the sides. (Make sure each side can be measured in whole inches.) Have students label each rectangle with the length of each side.

- **What is the length in inches of each side?**
- **Which side is the longest?**

Lesson Follow-Up

If students have difficulty completing this activity, have them complete Activity 23.

Activity 22

Emerging **Objective**

Students identify the sides on squares and rectangles.

Use pretzel sticks to make squares and rectangles. Explain that there are four sides on a square and rectangle. Model how to use the pretzels to make each side of the shape. Provide outlines of each shape for the students to place the pretzel sides on.

- **How many sides does a rectangle have?**
- **Are the sides of a rectangle equal?**
- **How many sides does a square have?**
- **Are the sides of a square equal?**

Lesson Follow-Up

If students have difficulty with this activity, have them use paper shapes and trace each side of the shape with a different color.

Conceptual Development Activities

Activity 23

Emerging Objective

Students compare rectangles and identify the differences in the sides of rectangles.

Cut out several rectangles of various sizes with the students. First, help students notice the differences in the length of the sides of a single rectangle.

- **This shape is called a rectangle. It has two long sides and two short sides.**
- **Show me/Tell me which sides of this rectangle are longer.**
- **Show me/Tell me which sides are shorter.**
- **How many sides in all?**

Next, have students compare different rectangles by placing the rectangles next to each other.

- **Which sides are the longest?**
- **Which sides are the shortest?**

Lesson Follow-Up

If students have difficulty with this activity, use a book and a calculator. Have students recognize the lengths of the sides of the book and the calculator by identifying the shortest and longest sides.

Activity 24

Independent Objective

Students identify, describe, and extend growing visual and number patterns using strategies, such as skip counting.

Tell students you are going to use counters to create a growing pattern. Place one counter in the first row, two counters in the second row, three counters in the third row, and four counters in the fourth row. Say:

- **Look at the counters in each row. Do you see a pattern?** Each row has one more counter than the one before.
- **How many counters do you think will be in the fifth row?**
- **What is the rule for this pattern?**

Next, have students identify and extend a growing number pattern by having them fill in the blanks. 5, 10, 15, ______, ______, 30, ______. Continue counting to a target number, such as 50.

- **What is the next number in this pattern?**
- **How do you know?**
- **What is the rule of the pattern?**

Finally, give students a 100 number grid. Have them use skip counting to count by twos, fives, and tens, and complete the pattern by coloring in the correct box.

Lesson Follow-Up

If students have difficulty completing this activity, have them complete Activity 27.

Conceptual Development Activities

Activity 25

Independent **Objective**

Students compare sets of equal and unequal objects and describe the sets using the following terms: greater than, less than, or equal to.

Give students two bags of items. Have students count the number of items in each bag and determine whether the number of objects in the bags makes equal or unequal sets. Have students use *greater than, less than,* and *equal to* vocabulary and symbols to describe each set of items.

- **Are the groups equal?**
- **Show me/Tell me how you know.**
- **Which group is greater than the other?**
- **Which group is less than the other?**

Repeat the activity using different amounts in each bag.

Lesson Follow-Up

If students have difficulty completing this activity, have them complete Activity 28.

Activity 26

Independent **Objective**

Students determine what rule, 1 less, 2 less, or 3 less, is represented in number pairs.

Draw number boxes on the board. Tell students they will match the number boxes with the rule: one less, two less, or three less. On the board, model several examples with the students. Say:

- **Here is a number pair (18, 17). What is the difference between the two numbers?** Give students a number line if they have trouble indicating the rule.
- **The rule is one less. How do you know?**

Continue with more number boxes and have students indicate 1 less, 2 less, or 3 less.

Sample number boxes are shown below.

35	33
19	17
6	4

- **What is the difference between the two numbers?**
- **What is the rule shown in the box?**
- **How do you know?**

Lesson Follow-Up

If students have difficulty with this activity, have them complete Activity 29.

Conceptual Development Activities

Activity 27

Developing **Objective**

Students create a two-element repeating pattern using objects and pictures.

Model for students how to make a two-element repeating pattern with letters or counters. Use the letters R and S to make a pattern with up to 20 letters. Next have students use the letters to create their own two-element repeating pattern.

- **How many different letters are there?**
- **What is the pattern?**

Have students create additional two-element repeating patterns using picture cards.

Lesson Follow-Up

If students have difficulty completing this activity, have them complete Activity 30.

Activity 28

Developing **Objective**

Students compare two sets of objects to 10 and determine if there is the same or different number in each set.

Give students two bags of up to 10 items each. Students can decorate the bags prior to the lesson. Have them count the number of items in each bag to determine if the number in the two sets is the same or different.

- **Are the groups the same?**
- **Are the groups different?**
- **Which group has more?**
- **Which group has less?**
- **Tell me how you know.**

Lesson Follow-Up

If students have difficulty completing this activity, have them complete Activity 31.

Conceptual Development Activities

Activity 29

Developing **Objective**

Students use a number line to determine 1 more or the next number with numbers 1 to 20.

Play a "One More" game. Create a set of index cards that have the numbers 0 to 19 written on them. Use a number line to review the numbers 0 to 20 with the students. Point out that the numbers are in line. Have students select an index card that has a number on it and then using the number line find the next number. For example, a student pulls 19 then asks, "What is the next number?" Point to the number 20. Continue using the number line to demonstrate identifying the next number. Then play the game One More by telling students a number and asking the students to tell you one more. "I have 16. What is one more? I have 15. What is one more?" Write the numbers on the board.

- **What is one more than 15?**
- **What is one more than 2?**
- **How do you know?**

Lesson Follow-Up

If students have difficulty completing this activity, have them complete Activity 32.

Activity 30

Emerging **Objective**

Students determine the next step in a pattern or sequence of activities.

Have students indicate the next step in a please-and-thank-you routine. Make picture cards paired with the matching word to respond to everyday greetings like *please* and *thank you*. Have students use the picture cards to indicate the step in the pattern. Another possible pattern is to use the hello-and-goodbye pattern in a sequence.

- **What do you say when you want something?**
- **What do you say when you get what you want?**

Lesson Follow-Up

If students have difficulty completing this activity, have them use shape cutouts such as circles and squares. Start the repeating pattern, and have students indicate which shape is next.

Conceptual Development Activities

Activity 31

Emerging Objective

Students compare sets of objects to 4 using one-to-one correspondence.

Distribute two sets of up to four crayons each. Have students use one-to-one correspondence to compare the sets, and determine whether the amount in each set is the same or different. Have students determine which set has more or less items.

- **Are the sets the same?**
- **How are the sets different?**
- **How do you know? Show me.**

Repeat the activity with different quantities up to 4.

Lesson Follow-Up

If students have difficulty completing this activity, have them use four-frame cards with large dots in each square. Use the dots to have students place a different amount of counters on each card. Then have students compare the amounts on each card.

Activity 32

Emerging Objective

Students add 1 more to a set of objects up to 3 and identify the quantity using one-to-one correspondence.

Play a "One More" game using objects. Give students 1 object. Place several additional items in a spot close to the students. Ask students to add one more object to their group by taking an item from the pile. Continue adding quantities up to 3 objects by adding one more. Have students count with one-to-one correspondence. Ask:

- **Show me/Tell me how many you have. Yes, you have 2. What is one more than 2?**
- **What is one more than 3?**
- **What is one more than 1?**

Lesson Follow-Up

If students have difficulty completing this activity, have them use smaller amounts of the items. Help them manipulate the items and count the items with one-to-one correspondence. Give them a number line up to 3. Have them point to the number to show one more.

Activity 33

Independent **Objective**

Students identify the angles on triangles and rectangles.

Explain that angles are the places where two sides of a shape join together. Have students use their arms to make angles. Then draw triangles on the board, and draw a curved line on each angle of the triangle. Draw additional triangles and rectangles on the board. Have students come to the board and locate the angles by drawing a curved line on each angle. Next, have students draw rectangles and triangles on a piece of paper and then locate the angles on each rectangle.

- **How many angles does a triangle have?**
- **Show me/Tell me where the angles are.**
- **How many angles does a rectangle have?**
- **Show me/Tell me where the angles are.**

Lesson Follow-Up

If students have difficulty completing this activity, have them complete Activity 36.

Activity 34

Independent **Objective**

Students identify symmetry and congruency using two-dimensional figures.

Remind students that congruent figures are the same size and shape. Display cutouts of a large rectangle and a small rectangle. Ask students if the shapes are congruent. (No.) Confirm by placing the small rectangle on top of the large rectangle. Then display a set of rectangles that are the same size and shape. Compare the rectangles, and determine if the rectangles are congruent. (Yes.)

Next, identify examples of figures that are visually the same on both sides of a central dividing line. Display a rectangle made of paper. Tell students: "Look at the rectangle. Let's fold it in half. Will it be the same on both sides?"

Finally, ask students where they see shapes that are the same shape and size in the environment. Where do they see shapes that are the same on both sides of a central dividing line?

- **Are these the same size and shape?**
- **Is this the same on both sides?**

Continue identifying more examples in the classroom and outside environment (books, desktops, and so on).

Lesson Follow-Up

If students have difficulty with this activity, have them complete Activity 37.

Conceptual Development Activities

Activity 35

Independent **Objective**

Students identify and sort three-dimensional objects into different categories.

Review the characteristics of three-dimensional objects. Display a group of three-dimensional objects on the table (basketball, block, cardboard tube, box, ice-cream cone). Have students sort the objects into various categories. Ask:

- **How many sides does this shape have?**
- **Are the sides straight or curved?**
- **Are the sides the same length?**
- **How many corners does this shape have?**

Encourage students to sort the shapes into different categories and explain their reasoning. Finally, help students name each shape (cube, cylinder, cone, rectangular prism, sphere).

Lesson Follow-Up

If students have difficulty completing this activity, have them complete Activity 38.

Activity 36

Developing **Objective**

Students identify the angles within a triangle.

Use 3 pretzel sticks to make a triangle. Have students point to each place where the pretzel sticks come together. Next have students use a highlighter to highlight the angles of paper triangles. Explain that angles are the places where two sides of the triangle join together.

- **What is an angle?**
- **How many angles are in a triangle?**
- **Show me/Tell me where the angles are.**

Lesson Follow-Up

If students have difficulty completing this activity, have them complete Activity 39.

Conceptual Development Activities

Activity 37

Developing **Objective**

Students fold two-dimensional figures to identify symmetrical shapes.

Have students cut out various shapes such as hearts, circles, diamonds, and squares. Have students fold each shape in half. Then have students identify whether the shape is the same on both sides of the fold/dividing line. Continue by looking at some examples of irregular 3, 4, and 5 sided figures that have no line of symmetry.

- **Here is a shape. Is the shape the same on both sides?**
- **How do you know?**

Lesson Follow-Up

If students have difficulty completing this activity, have them complete Activity 40.

Activity 38

Developing **Objective**

Students identify the characteristics of three-dimensional objects and match the objects with three-dimensional shapes.

Review the characteristics of three-dimensional objects including number of sides, whether sides are straight or curved, and so on. Show students models of each three-dimensional shape (cube, cylinder, cone, and sphere).

Next, display several classroom objects, such as a ball, a construction cone, a basketball, a cereal box, and a toilet-paper tube. Have students match each object with the correct model. Identify objects that don't belong, such as rectangular prisms (cereal box). Ask:

- **Show me/Tell me which object is a sphere. How do you know?**
- **Which object is a cone? How do you know?**

Lesson Follow-Up

If students have difficulty completing this activity, have them complete Activity 41.

Conceptual Development Activities

Activity 39

Emerging **Objective**

Students identify the number of corners or angles on squares and rectangles.

Collect a paper bag and cardboard squares and rectangles. Hide the squares and rectangles in the bag. Have the students put their hands into the bag without looking inside and feel the corners on each shape. Ask students how many corners they feel, and record the guess each student makes. If students are unable to complete this task then modify this task by placing the shape in students' hands while keeping the shape covered with your hand so students cannot see it. Students also can use number cards to show the number of corners. Take each shape out of the bag. Work with students to identify and count the corners of the shape. Continue until every student has a turn.

- **Here is a square. Show me/Tell me where the angles are.**
- **Here is a rectangle. Show me/Tell me where the angles are.**

Lesson Follow-Up

If students have difficulty with this activity, have them search the classroom for objects that have four corners, like a square or a rectangle. Give students a cardboard square and a cardboard rectangle to refer to as they search for squares and rectangles in the room.

Activity 40

Emerging **Objective**

Students compare the two sides of two-dimensional figures to identify symmetrical shapes.

Give students several shapes that are already cut out and folded on the central dividing line. Have students use the shapes to recognize each side of the symmetrical shape. Include shapes that are not symmetrical. Have students sort the symmetrical versus the non-symmetrical shapes.

- **Here is a shape. Show me/Tell me one side of the shape.**
- **Is the shape the same on both sides?**

Continue by showing students one side of a particular shape and have the students choose the other side from a selected group of choices.

Lesson Follow-Up

If students have difficulty with this activity, have them use alphabet letters to recognize the letters that are symmetrical.

Conceptual Development Activities

Activity 41

Emerging **Objective**

Students identify three-dimensional objects.

Review common names of three-dimensional objects (ball, block, tube) with students, and display models of the objects. Then show students a selection of classroom objects. Guide students through the process of using the models to recognize where each classroom object belongs. For example, ask, "Which object does the basketball look like? Is the basketball a sphere or a cube?" Guide students with questions such as:

- **Which shape does this look like?**
- **Does it have flat sides or round sides?**
- **Does it have corners?**
- **What is the name of this shape?**

Lesson Follow-Up

If students have difficulty with this activity, have them match pictures with the three categories (cube, sphere, cylinder).

Activity 42

Independent **Objective**

Students represent whole numbers to 50 in various contexts.

Give students a box with a number between 0 and 50 written on the outside. Model expressing and representing the number in different ways. For example, say "What are six names for 25? $24 + 1 = 25$, one quarter = 25, 5 nickels = 25, $5 \times 5 = 25$, $20 + 5 = 25$, $30 - 5 = 25$." Give students a number collection box, and have them express or represent a number with at least five names.

- **What addition or subtraction fact represents this number?**
- **Which coins do you need to express this number?**
- **How many sets of 5 represent this number?**
- **How do you write this number?**

Repeat the activity with various numbers from 0 to 50.

Lesson Follow-Up

If the students have difficulty completing this activity, have them complete Activities 47 and 48.

Conceptual Development Activities

Activity 43

Independent **Objective**

Students solve addition and subtraction problems using related facts.

Draw fact-family triangles on the board, and review fact families with students. For example, draw a triangle with 7, 9, and 16 at the corners. Work as a class to list all the ways these numbers fit together, and write the equations on the board.

$$7 + 9 = 16$$
$$9 + 7 = 16$$
$$16 - 7 = 9$$
$$16 - 9 = 7$$

Repeat with several fact families. Then give students addition problems and related subtraction problems to solve. Help them use fact families and the inverse relationship of addition and subtraction when solving the problems.

For example: "I brought 6 apples to a picnic. My friend brought 5 apples. How many apples were there altogether?" ($6 + 5 = 11$)

"We had 11 apples. We ate 5 apples. How many apples were left?" ($11 - 5 = 6$)

As students solve the problems, ask:

- **What numbers are in the problem?**
- **What is the addition problem?**
- **What is the related subtraction problem?**
- **How do you know?**

Lesson Follow-Up

If the students have difficulty completing this activity, have them complete Activity 49.

Activity 44

Independent **Objective**

Students identify the relationship between whole, halves, and fourths.

Give students shapes to fold into halves and fourths, and to leave as a whole. For example, give students three paper squares that are the same size. Have them fold one paper square into fourths and color each fourth a different color. Have students fold the second square in half and color each half a different color. Have students color the last square entirely. Have students identify the shapes by halves, fourths, and a whole.

- **Which shape is shaded in fourths?**
- **Which shape is shaded in half?**

Then have students compare the fourths, halves, and whole to one another.

- **How many fourths make up a whole?**
- **How many fourths make up a half?**
- **How many halves make up a whole?**

Lesson Follow-Up

If students have difficulty completing this activity, have them complete Activity 50.

Activity 45

Independent **Objective**

Students skip count by 5s and 10s up to 50.

Give students a handful of nickels and dimes. Write an amount on the board. Model skip counting from 0 to 25 using the nickels (5, 10, 15, 20, 25). Give students various amounts to count from to 50. Let them use the nickels for skip counting by 5s and the dimes for skip counting by 10s.

- **Start from 25 and count by 5s to 50.**
- **Start from 20 and count by 10s to 50.**

Alternatively, give each student several number lines from 0 to 50. Have students use a pencil/marker to make hops from each number. Have students practice skip counting by 5s and 10s.

Lesson Follow-Up

If students have difficulty with this activity, have them complete Activity 51.

Activity 46

Independent **Objective**

Students compare and estimate groupings of objects to 20.

Display a set of 20 play one-dollar bills. Then, give students a set of 9 play one-dollar bills and a second set of 3 play one-dollar bills. Put 9 play one-dollar bills into a group, and put 3 play one-dollar bills into a group. Count how many dollar bills you have in each group. Have students compare the groups of dollar bills to the display of 20 one-dollar bills to estimate to 20.

- **Which set of dollar bills is closer to 10?**
- **Which set of dollar bills is closer to 1?**

Repeat the activity with various quantities up to 20.

Lesson Follow-Up

If students have difficulty with this activity, have them complete Activity 52.

Activity 47

Developing **Objective**

Students represent whole numbers to 25 using objects, pictures, number names, and numerals.

Give students a box with a number between 0 and 25 written on the outside. Collect straws, paper coins, craft sticks, picture cards showing sets from 0 to 25, and counters that students can use with the number box. For example, tell students: "My number box is 10. I am going to put 10 straws, 10 counters, 2 nickels, and 1 dime in the number box to each represent the number 10." As students fill their number boxes, ask:

- **What is the name of the number on your box?**
- **How do you write the number?**
- **How many craft sticks make the number?**
- **Which picture card matches your number?**

Repeat the activity with various numbers to 25.

Lesson Follow-Up

If the students have difficulty completing this activity, have them complete Activity 53.

Activity 48

Developing **Objective**

Students use ordinal numbers to describe real life situations.

Ask students to form a line in the classroom. Use ordinal numbers to describe the students' sequence from first to last. For example, say: "Gina is first. Theo is second." Show students the number cards with the numeral, number name, and ordinal name on the card to reinforce the ordinal numbers.

Have students form a new line. Ask:

- **Who is first?**
- **Who is second?**
- **Who is third?**
- **Who is last?**

Have students put the ordinal numbers in sequence to match how the students are lined up.

Lesson Follow-Up

If students have difficulty with this activity, have them complete Activity 53.

Activity 49

Developing **Objective**

Students represent addition facts and related subtraction facts using objects and pictures.

Distribute 15 pennies and a sheet of paper to each student. Use the pennies to make addition and subtraction facts. Say, "I have 3 pennies, and I added 4 more pennies, which equals 7 pennies. $3 + 4 = 7$." Continue by subtracting. Say, "I have 7 pennies, and I took 4 away. I have 3 pennies. $7 - 4 = 3$." Have students repeat the process and write as many addition and subtraction facts up to 15 as they can.

- **What is the addition fact?**
- **What is the related subtraction fact?**

Have students repeat the activity using picture cards.

Lesson Follow-Up

If the students have difficulty completing this activity, have them complete Activity 54.

Activity 50

Developing **Objective**

Students identify the relationship between whole and half using liquid measurements.

Use measuring tools such as 1 teaspoon, $\frac{1}{2}$ teaspoon, 1 cup, and $\frac{1}{2}$ cup. Have students use the measuring tools to fill a small, clear container with water. Start with the $\frac{1}{2}$ teaspoon or $\frac{1}{2}$ cup, and draw a black line to show the water level. Empty the water, and fill the container again using the cup and teaspoon. Draw a black line to show the water level for each of these. Discuss with students the relationships between half and whole.

- **What happened to the line of water when 1 cup was added?**
- **What is the relationship between 1 cup and $\frac{1}{2}$ cup?**
- **Which has more water, $\frac{1}{2}$ teaspoon or 1 teaspoon?**

You may also have students use the $\frac{1}{2}$ cup measure to fill the cup. Ask:

- **How many half cups are in a cup?**
- **How many halves are in a whole?**

Lesson Follow-Up

If students have difficulty completing this activity, have them complete Activity 55.

Activity 51

Developing **Objective**

Students create equal sets using objects to 25 and name the total number of objects in each set and the total number of sets.

Give students 12 pennies or counters. Have them separate the counters into groups of 4. Have students identify the total number of sets (3). Repeat with 25 pennies or counters. Have students separate the pennies into groups of 5. Have students identify the total number of sets (5) and identify the number of pennies it took to make each set (5). Write the number model on the board ($\frac{25}{5}$ or 5×5). Give students various amounts of pennies to separate into equal sets (20, 15, 10, 14, 6, 2). Write the number model on the board.

Lesson Follow-Up

If students have difficulty completing this activity, have them complete Activity 54.

Activity 52

Developing **Objective**

Students compare and estimate groupings of objects to 10.

Give students a set of pennies. Have students separate the pennies into groups. Put 4 pennies into a group, and put 8 pennies into another group. Count how many pennies are in each group. Have students compare the groups of pennies to estimate to 10.

- **Which set of pennies is closer to 10?**
- **Which set of pennies is closer to 5?**

Lesson Follow-Up

If students have difficulty with this activity, have them complete Activity 53.

Conceptual Development Activities

Activity 53

Emerging **Objective**

Students represent quantities to 4 using objects, pictures, or number names.

Give students pictures of items showing quantities from 1 to 4, small objects to manipulate, and number name cards. Share a number name card with the students. Have students use the objects and pictures to make number collections by placing the items together. For example, with the number name 4, students put 4 pennies and 4 cat pictures in their collection.

- **How many pictures or items do you need to show 4?**
- **Tell me how many items you need to show this number.**

Continue with the numbers 0, 1, 2, 3, and 4.

Lesson Follow-Up

If students have difficulty with this activity, have them work with fewer items. Give them the number card and a set of 2 objects. Have them match two objects with the number name.

Activity 54

Emerging **Objective**

Students create equal sets of objects using groups of objects to 4.

Give students a deck of cards with numbers 2, 3, and 4. There should be four of each number. Have students separate the deck of cards into equal groups of four. Have students put all the 2 cards into a set, and put the 3 cards into a set and all the 4 cards into a set. Have students count with one-to-one correspondence the total number of cards in each set.

- **Show me/Tell me how many cards are in each set.**
- **Are all the sets the same size?**

Lesson Follow-Up

If students have difficulty completing this activity, give students 8 colored counters (4 red and 4 blue). Have students separate the counters into red and blue groups of 4.

Conceptual Development Activities

Activity 55

Emerging **Objective**

Students match parts to whole objects.

Display a box of crayons. Have a student take a red crayon. Discuss that the red crayon is part of the whole box of crayons. Have each student take a turn choosing a different crayon from the box. Each time discuss that the crayon is part of the whole set of crayons. Continue having students match different crayons with the crayon box.

- **Show me the whole box of crayons.**
- **Show me the red crayon.**
- **Is the red crayon part of the whole box of crayons?**

Lesson Follow-Up

If students have difficulty completing this activity, use picture examples: a piece of cake is part of the whole cake, a slice of pizza is part of the whole pizza, or an orange slice is part of the whole orange.

NOTES

5

Conceptual Development Activities

Activity 1

Independent Objective

Students informally divide by separating objects into equal sets.

Demonstrate *division* by having students group objects, such as coins, counters, or dominoes, into smaller but equal groups. Begin with a group of 20 counters and 5 boxes or plates. Have students count all the counters and then begin placing them one at a time into a box or on a plate until all the counters are gone. Ask,

- **Which had the greater amount, the whole group of objects or the divided groups?**
- **How many objects are in the larger group?**
- **How many objects are in each of 5 boxes?**

Repeat this activity using 30, 40, and 50 objects. Possible ideas include coins, cards, or base-ten blocks. Ask students to divide the groups equally into smaller groups of 6, 8, and 10. Ask students to do the following:

- **Divide the 30 cards into five equal groups. How many cards do you have in each group?**
- **Divide the 40 coins into five equal groups. How many coins are in each group?**
- **Divide the 50 base-ten blocks into five equal groups. How many base-ten blocks are in each group?**

Repeat this activity by grouping various objects into equal groups. Have students see which numbers can and cannot be equally divided.

Lesson Follow-Up

If students have difficulty completing this activity, have them complete Activity 3.

Activity 2

Independent Objective

Students model multiplication and division using up to 50 objects.

Demonstrate for students a 5-column, 4-row array of cards on a table. Have students count the 20 cards one at a time to see how many there are. Show them $4 \times 5 = 20$. Collect the cards, and distribute them among 4 different boxes. Have students count the number of cards in each box. Show them $20 \div 4 = 5$. Discuss how 4 groups of five equal 20. Confirm by counting the cards. Repeat this activity for 25 and 40 objects. Finally, display a set of 50 objects, such as marbles, coins, or picture cards. Say,

- **Here is a set of 50 cards. These cards can be divided equally into smaller sets. Divide the cards into equal sets of 5.**
- **Show me/Tell me into how many sets you divided the cards.**
- **Show me/Tell me how many cards are in each set.**

Write (or ask students to write) a division problem that represents the above scenario (for example, 5 sets of 10 cards). Demonstrate how to use multiplication to check your work.

Lesson Follow-Up

If students have difficulty completing this activity, have them complete Activity 4.

Conceptual Development Activities

Activity 3

Developing Objective

Students use objects and pictures to count and group quantities to 25 into equal sets.

Discuss the term *divide* by demonstrating how groups of objects, such as building blocks or coins, can be divided into smaller but equal groups. Show students 4 objects such as coins or crayons. Ask questions such as

- **Count the objects in the whole group. How many objects are in this group?**

Set 2 crayons on one plate and 2 crayons on a second plate.

- **Count the objects in each of the divided groups. How many objects are in each of these groups?**

Place a grouping of 25 objects before the students. Ideas include 25 pencils or 25 bills of play money.

- **Here is a group of pencils. We are going to divide the large group into smaller groups**

Work together to divide the pencils into equal groups by placing them onto plates until the pencils are gone. Ask the following questions:

- **How many groups of pencils do we have?**
- **How many pencils are in each group?**

Repeat by dividing the bills into equal groups. Ask,

- **How many groups of bills do we have?**
- **How many bills are in each group?**

Lesson Follow-Up

If students have difficulty completing this activity, have them complete the Activity 5.

Activity 4

Developing Objective

Students combine (multiply) or separate (divide) equal sets of objects and pictures up to 25.

Review how the terms *multiplying/combining* and *dividing/separating* are related. Display a set of 10 objects, such as base-ten blocks or coins.

- **Here is a grouping of 10 coins.**

Show students how to divide the coins into 2 equal groups of 5. Ask students:

- **Show me/Tell me into how many smaller sets was the group divided.**
- **Show me/Tell me how many coins are in each set.**

Repeat the above activity with 15, 20, and 25 objects. Display the items in arrays of 3×5, 4×5, and 5×5. Have students count either one at a time or skip count by 5s to confirm the total number of objects. Then have students divide the objects into equal groups of 5; for example, 3 sets of 5 coins. Finally, explain how combining the sets will lead you back to the total number of coins from the original group.

Lesson Follow-Up

If students have difficulty completing this activity, have them complete Activity 6.

Activity 5

Emerging **Objective**

Students informally divide by grouping objects up to 4 into equal sets and identifying how many are in each set.

Display common classroom objects in groupings of 4 (books or pencils, for example).

- **Here is a grouping of books. It is one large group. Tell me how many groups of books you see.**
- **Show me/Tell me how many books are in this group.**

Separate the books into two smaller sets of two. Explain that the large group has been separated into two groups of two. Continue by asking students:

- **Show me one of the smaller groups of books.**
- **Show me/Tell me how many books are in this group.**

Repeat this activity with different objects in groups of four or less.

Lesson Follow-Up

If students have difficulty completing this activity, have them repeat it with different classroom objects on several occasions until students are comfortable with the routine.

Activity 6

Emerging **Objective**

Students join and separate objects to 5 to solve simple problems.

Display common classroom objects in groupings of 5 (counters, coins, or craft sticks, for example). Say:

- **Here is one group of counters. This group contains 5 counters.**

Continue by asking students:

- **Show me/Tell me how many groups of counters you see.**
- **Count how many counters are in this group.**
- **We can separate these counters into smaller sets.**

Place three counters in one group and two counters in another group. Ask students:

- **Show me/Tell me which group has 3 counters.**
- **Show me/Tell me which group has 2 counters.**

Demonstrate that the counters can then be joined together again. Place the three counters with the two counters, and count all five counters.

- **Show me/Tell me what happens when we put all the counters back together in one group. Show me/Tell me how many counters we have.**

Lesson Follow-Up

If students have difficulty completing this activity, repeat the above activity with different classroom objects or different groupings of two, three, or four objects on several occasions until students are comfortable with the routine.

Conceptual Development Activities

Activity 7

Independent Objective

Students represent fractions including halves, fourths, and thirds as parts of a set and parts of a whole.

Demonstrate a fraction by taking a sheet of paper, folding it in half, and shading half. Explain that half or 1 out of 2 possible areas are shaded. Write the fraction $\frac{1}{2}$ on the board. Show a group of 4 pieces of candy. Say:

- **Here is a group of four candies. We can show how many candies are in this one group by writing $\frac{4}{4}$. I am going to eat two of the candies. Because I ate two candies, we now have two out of four candies left.**

Write the fraction $\frac{2}{4}$ on the board.

- **If I ate one more candy, what fraction would be left?**

Explain that $\frac{1}{4}$ is left. Repeat the above activity by having students express the fraction of candies they have. Ask students:

- **Show me $\frac{1}{3}$ of your candies.**

Continue the activity by showing students index cards with the fractions $\frac{1}{2}$, $\frac{1}{4}$, and $\frac{1}{3}$ written on them. Fold a sheet of paper, and shade a section to reflect a particular fractional amount written on the index cards. Have students show you or tell you which fraction of the paper is shaded.

Lesson Follow-Up

If students have difficulty completing this activity, have them complete Activity 12.

Activity 8

Independent Objective

Students model and express whole numbers up to 100.

Set 5 counters on the table, and have students count the items. Tell students that when they count they use *whole numbers*. Write several whole numbers that are less than 100 on the board. Using base-ten blocks, demonstrate a couple of the numbers on the board. Point out to students the tens and ones places from the numerals, and show them how that translates to the base-ten blocks. Continue by displaying various numerals between 1 and 99 on a set of index cards. Have students model the number using the base-ten blocks. Then model a number, and have the students select the index card showing the correct numeral. Ask questions such as

- **If we have 23 pieces of chalk, how is 23 represented?**
- **If we have 77 craft sticks, how is 77 represented?**

Have the student either write, say, or model these numbers. Cover other whole numbers using similar processes. End the activity by modeling 100 using base-ten blocks. Write the number 100, and point out the hundreds column in the number. Work with students to identify items in the classroom that could be equivalent to 100.

Lesson Follow-Up

If students have difficulty completing this activity, have them complete Activity 13.

Activity 9

Independent Objective

Students compare fractional parts of objects, including halves, fourths, and thirds.

Review the terms *fraction, fractional parts,* and *whole*. Remind students that a fraction is part of a whole or set. Hold up a ruler. Explain that a ruler is one foot in length. One foot is made up of 12 inches. The inches are parts, or fractions, of one whole foot. Pass the ruler around to all students, or have students hold their own rulers. Ask:

- **One ruler is equal to how many feet?** 1
- **Show me a fraction of one foot.**
- **What do you call a fraction of a foot?** Inches

Review the terms *halves, fourths,* and *thirds*. Explain that half of one foot, or 12 inches, is six inches. Ask students to show you what six inches look like on the ruler. Do the same for one fourth and one third of a ruler. Also consider doing similar activities involving paper folding or the manipulation of learning links or connecting cubes.

Lesson Follow-Up

If students have difficulty completing this activity, have them complete Activity 14.

Activity 10

Independent Objective

Students identify place value of numbers up to 99 in terms of tens and ones.

Review the terms *two-digit numbers* and *place value*. Write various two-digit numbers from 10 to 99 on the board or on index cards. Point to and explain the tens position and the ones position on several of the numbers. Continue by asking students:

- **Come to the board, and point to one two-digit number.**
- **Show me/tell me the number in the tens position.**
- **Show me/tell me the number in the ones position.**

Continue the activity with more volunteers. Then ask students to write their own two-digit numbers on paper or index cards. Ask volunteers to

- **Hold up your numbers. Show the class the number in the tens position.**
- **Show the class the number in the ones position.**

Draw a two-column chart on the board. Label the ones and the tens positions. Ask volunteers to come to the board and fill in the chart with their numbers.

Lesson Follow-Up

If students have difficulty completing this activity, have them complete Activity 15.

Conceptual Development Activities

Activity 11

Independent **Objective**

Students compare fractions, including halves, fourths, and thirds, using parts of objects of equal size.

Use fraction circles to express and represent fractions, including halves, thirds, and fourths. Give students paper fraction circles cut into halves, thirds, and fourths. Have students assemble the pieces to create a whole and then label each part of the circle with the appropriate fraction. Ask students to refer to the pieces as they answer the following questions:

- **Look at $\frac{1}{2}$ and $\frac{1}{4}$. Show me the larger fraction.**
- **Look at $\frac{1}{2}$ and $\frac{1}{3}$. Show me the larger fraction.**
- **Look at $\frac{1}{3}$ and $\frac{1}{4}$. Show me the larger fraction.**

Repeat this activity by folding sheets of paper and shading fractional equivalents to $\frac{1}{2}$ and $\frac{1}{4}$ of the paper. Do this to a second sheet of paper, but fold the paper into thirds, then shade one of the sections.

Lesson Follow-Up

If students have difficulty completing this activity, have them complete Activity 14.

Activity 12

Developing **Objective**

Students represent fractions, such as halves and fourths, as parts of a whole.

Discuss the term *fractions* with students by demonstrating how to share equally. Show them a clear glass of water.

- **I would like to share this water with you. As I pour, show me/tell me when you and I have the same amount.**

Pour the water from your cup into a clear cup. When students tell you to stop pouring, confirm that the level of water in the cups is equal. Explain that there was a whole glass of water, but now you both have $\frac{1}{2}$ the amount of water. Next, create a circle with two sections on the board. Say,

- **Here is a circle with two sections. I want to shade in half of the circle. Show me/Tell me when half the circle is shaded.**

When students tell you to stop, confirm that half is shaded. Write $\frac{1}{2}$ on the board; explain that the fraction names what is shaded. Repeat the circle activity with fourths. Ask:

- **Show me/Tell me when $\frac{1}{4}$ is shaded.**
- **Would you say that two of the four sections are one-half or one-fourth of the whole?**

Confirm students' responses by comparing the shaded areas of the circles.

Lesson Follow-Up

If students have difficulty completing this activity, have them complete Activity 16.

Conceptual Development Activities

Activity 13

Developing **Objective**

Students use whole numbers up to 30 and ordinal numbers up to fifth.

Review the term *whole numbers* with students. Write the whole numbers 1 to 30 on the board. Explain that they can count with whole numbers. Count with students from 1 to 30. Point to various numbers on the board, and have students model these numbers using base-ten blocks or one- and ten-dollar bills of play money. Point out the tens and ones places of the numbers on the board and the corresponding tens and ones places of the base-ten materials. Next write several numbers on index cards. Create one of the numbers using base-ten materials.

- **Show me/Tell me what number is shown.**

Repeat this several more times with various numbers. Make sure that students do not confuse the digits in the tens and ones places. Finally, discuss *ordinal numbers*. Place five objects on a desk, or draw five objects on the board. Explain the first, second, third, fourth, and fifth positions.

- **Show me/Tell me which object is in first position.**

Continue by having students name or show the objects in second through fifth positions.

Lesson Follow-Up

If students have difficulty completing this activity, have them complete Activity 17.

Activity 14

Developing **Objective**

Students understand fractions, including fourths, halves, and wholes.

Review *fractional parts* and *whole* by cutting shapes out of paper and folding them in half. For example, cut out a circle, and have students fold the shape in half. Shade one half of the circle, and talk about how the circle is whole and $\frac{1}{2}$ the circle is shaded. Continue with other shapes. Ask the following:

- **Show me the whole rectangle. Fold it in half, and shade $\frac{1}{2}$.**

Repeat, but fold shapes into fourths. Have students compare the fractional sizes of the folded shapes. Explain how halves are two equal pieces of a whole and fourths are four equal pieces of a whole. Select a particular shape, and have students compare the sizes of whole to half and half to fourth by answering the following:

- **Show me/Tell me which circle is whole. Is this whole circle bigger than this half of a circle?**
- **Show me/Tell me which is smaller— $\frac{1}{2}$ or $\frac{1}{4}$ of a rectangle.**

Lesson Follow-Up

If students have difficulty completing this activity, have them complete Activity 18.

Conceptual Development Activities

Activity 15

Developing **Objective**

Students count and group by tens and ones to identify numbers up to 30.

Review the terms *two-digit numbers* and *place value.* Write numbers from 10 to 30 on the board. As a class, count the numbers aloud. Give students base-ten materials such as base-ten blocks or one- and ten-dollar bills of play money. Have students count again from 1 to 10, making trades of one-dollar bills for ten-dollar bills when appropriate. Repeat again for numbers 11 to 20 and 21 to 30. Draw a two-column chart on the board. Label the ones and tens place values. Select a number such as 17. Model the number using the base-ten materials, show the number, and then show how the number is placed on the place value chart. Model a different number, and ask:

- **I've modeled a number. Show me/Tell me what number I've modeled.**
- **Show me/Tell me which number goes in the ones place. Which goes in the tens place?**

Repeat with various numbers. Finally, write some numbers between 1 and 30 on index cards. Show the card to the students, and ask:

- **Show me/Tell me the number in the tens place.**
- **Show me/Tell me the number in the ones place.**

Lesson Follow-Up

If students have difficulty completing this activity, have them complete Activity 18.

Activity 16

Emerging **Objective**

Students use sets of objects and whole objects to identify parts of a whole.

Display a group of four objects, such as four books in a set. Say:

- **Here is a set of four books.**

Pick up one book, pretend to quickly read it, and say:

- **I read 1 book. I read a part of the whole set.**

Continue by picking up the rest of the books one at a time and pretending to read them. After you have picked up all the books, say:

- **I read the whole set of books.**

Place the four books back on the table. Ask:

- **Show me/Tell me which is 1 part of the set.**
- **Show me/Tell me which are 2 parts of the set.**

Display an object that can be divided into pieces, such as a large jigsaw puzzle. Say:

- **Here is a whole puzzle. This puzzle has 6 pieces.**

Remove a piece of the puzzle, and say:

- **This is 1 piece of the puzzle. Show me another piece of the puzzle.**

Repeat, removing more puzzle pieces. Ask:

- **Show me/Tell me how many pieces of the puzzle I have.**

Put the puzzle together again. Ask:

- **Is this a whole puzzle or 1 piece of the puzzle?**

Lesson Follow-Up

If students have difficulty completing this activity, have students look at a small grouping of objects, such as a group of six pencils. Ask students to point to or pick up one of the objects. Then ask students to point to or pick up two of the six objects. Repeat the activity with different objects.

Conceptual Development Activities

Activity 17

Emerging **Objective**

Students use objects and visual models to identify half and whole.

Review the words *half* and *whole*. Hold up an orange. Say, "This is a whole orange. I want to eat just half of this orange." Cut the orange in half, and hold up half of the orange. Say, "I cut the whole orange into two halves or 2 parts. This is one half of the whole orange." Hold the two pieces together, and ask:

- **Show me/Tell me if you have a whole or a half when you put the halves of the orange together.**
- **Show me/Tell me how you make a whole orange into two halves.**

Give students sheets of notebook paper. Demonstrate tearing a sheet of paper in half. Continue by asking students to do the following:

- **Hold up your sheet of paper.**
- **Show me/Tell me if you have one whole piece of paper or one half piece of paper.**
- **Tear your sheet of paper into two pieces.**
- **Tell me if you have a whole piece of paper or two halves.**

Lesson Follow-Up

If students have difficulty completing this activity, continue to model whole objects that can be broken into halves. If possible, have students hold the objects while you talk about them.

Activity 18

Emerging **Objective**

Students compare up to 5 objects.

Display three separate sets of objects (pencils, marbles, and books, for example). Have five objects in two of the sets and three objects in the third set. Say, "Here are three sets of objects. I'm going to count the objects in each of the three sets." Say, "Here is a set of pencils. There are 1, 2, 3, 4, 5 pencils in this set." Continue with the other two sets, and then say to students:

- **Point to a set with five objects.**
- **Point to another set with five objects.**
- **Point to a set with three objects.**

Review the sets once again with students. Ask:

- **Show me/Tell me if all the sets have the same quantities.**

Ask students to count the objects in each of the sets.

- **How many objects are in the first set? The second set? The third set?**

Lesson Follow-Up

If students have difficulty completing this activity, place them into groups of various sizes, some same and some different. If possible, have students count off numbers. Have a volunteer in each group tell how many are in their group. Compare groups.

Conceptual Development Activities

Activity 19

Independent Objective

Students identify edges, sides, and faces of three-dimensional solids.

Draw the following objects on the board: a square, a triangle, and a circle. Label and say aloud the names of the different shapes. Explain similarities and differences between the shapes. For instance, explain that all of the shapes are two-dimensional because they do not have depth. Explain that they are different because they have different numbers of edges (or no edges), and some have curved and some have straight lines. Ask students:

- **Tell me how many edges the square has.**
- **Tell me the number of edges the triangle has.**
- **Tell me which shape has curved lines.**
- **Tell me which shapes have straight lines.**

Explain three-dimensional solids. Show students examples of three-dimensional real-world objects. Examples could include a ball, a box, or a cup. Draw a cube, a cone, and a sphere on the board. Ask students:

- **Show me/Tell me which solid looks most like a square. Which looks most like a triangle? A circle?**
- **Show me/Tell me what the difference is between the shapes and the solids.**

Lesson Follow-Up

If students have difficulty completing this activity, have them complete Activity 21.

Activity 20

Independent Objective

Students identify faces, edges, and corners on three-dimensional objects and models.

Review the term *three-dimensional*. Display three-dimensional rectangular prism-shaped classroom objects (such as the top of a desk or a bookcase). Point to the objects and review the six faces of the objects. Continue by asking students:

- **Tell me why these objects are called three-dimensional.**
- **Show me how many faces each of these objects has.**
- **Tell me how many corners each of these objects has.**

Challenge students to identify other objects in the classroom that are three-dimensional rectangular prism-shaped. Ask students to bring these objects to the front of the classroom and demonstrate how and why the objects fall into this category. Support students as needed. Ask:

- **Show us the six faces of your object.**
- **Tell us how many corners your object has.**
- **Count the number of edges your object has.**

Lesson Follow-Up

If students have difficulty completing this activity, have them complete Activity 22.

Conceptual Development Activities

Activity 21

Developing **Objective**

Students identify edges, sides, and corners in two- and three-dimensional shapes.

Draw the following shapes on the board: a square, a triangle, and a circle. Label and say out loud the names of the different shapes. Ask students to repeat the names of the shapes with you. Ask,

- **How many edges does the square have? How many corners?**
- **How many edges does the triangle have? How many corners?**
- **Tell me if the circle has curved or straight sides.**

Display three-dimensional classroom objects: a cube, a cone, and a sphere (for example a hamster cage or an aquarium, an ice-cream cone, and a globe). Ask students:

- **Tell me which object looks most like the square.**
- **Tell me which object looks most like the triangle.**
- **Tell me which object looks most like the circle.**

Ask students to count the number of corners in the cage and the ice-cream cone. Ask them if they can identify a shape other than a cone in the ice-cream cone.

Lesson Follow-Up

If students have difficulty completing this activity, then have them complete Activity 23.

Activity 22

Developing **Objective**

Students identify the faces of a three-dimensional object.

Review the term *three-dimensional*. Display three-dimensional rectangular prism-shaped classroom objects (such as an eraser, a book, or a pencil box). Hold up the objects, and review the six faces of the objects. Say, "Here is a three-dimensional prism-shaped object. It has six faces." Point out all six faces. Have volunteers hold the objects. Ask students,

- **Show me the six faces on the object.**
- **Count the number of faces on the object.**
- **Tell me how many faces the object has.**

Locate additional rectangular prism-shaped objects in the classroom (or bring in other objects or pictures of objects). Show students the objects or pictures and tell students that the objects are three-dimensional. Ask students,

- **Show me/Tell me the faces of the object.**
- **Show me/Tell me how many faces the object has.**

Lesson Follow-Up

If students have difficulty completing this activity, then have them complete Activity 24.

Conceptual Development Activities

Activity 23

Emerging **Objective**

Students identify and name the differences between two- or three-dimensional objects.

Draw the following shapes on the board: a square, a triangle, and a circle. Label and say aloud the names of the different shapes as you trace the shapes with your finger. Ask students to repeat the names of the shapes with you. Retrace the shapes a second time, this time counting the number of corners on the square and the triangle. For the circle say, "This shape does not have corners; it is one curved line." Provide students with index cards of the images. Ask:

- **Point to the shape with four corners. Trace the shape with your finger.**
- **Point to the shape with three corners. Trace the shape with your finger.**
- **Point to the shape with one curved line. Trace the shape with your finger.**
- **Show me/Tell me which shape is a square.**
- **Show me/Tell me which shape is a triangle.**
- **Show me/Tell me which shape is a circle.**

Lesson Follow-Up

If students have difficulty completing this activity, throughout the day periodically point out various objects with these shapes. Encourage students to hold the objects and trace their shapes.

Activity 24

Emerging **Objective**

Students identify size differences in two- and three-dimensional objects.

Hold up an image of a square drawn on an index card. Tell students squares are two-dimensional and cannot be picked up. Hold up a pencil or a tissue box. Tell students that prisms are three-dimensional and can be picked up. Illustrate various sizes of two-dimensional objects on index cards. For example, on one card show a square that is 1 inch by 1 inch, and on a second card show a square that is 3 inches by 3 inches. Point out the differences in sizes of the squares. Reinforce this by having students trace the squares with their fingers. Continue with other figures, such as circles, triangles, and pentagons. Ask:

- **Are these figures the same?**
- **Show me/Tell me which figure is larger. Which is smaller?**

Next, take some three-dimensional objects—for example, self-sticking notes, sheets of paper, pencil boxes, and canned foods—and show them to the students. Have students handle and trace the edges of the figures. Collect the objects, and display two of the same shape, such as two rectangular prisms. Ask:

- **Are these figures the same?**
- **Show me/Tell me which figure is larger. Which is smaller?**

Lesson Follow-Up

If students have difficulty with this activity, cut out the two-dimensional shapes for the students and place them on top of each other.

Conceptual Development Activities

Activity 25

Independent **Objective**

Students solve problems by maintaining equality.

Review the term *equality* with students. Write the equation $2 = 2$ on the board. Say, "If you add or subtract the same number to each side of this equation, the sides remain equal." Use a balance scale to demonstrate the concept of equality. Place individual base-ten blocks on each side of the scale. Explain that the blocks are like numbers, and when the scale is balanced, both sides are equal. Draw a picture of a seesaw on the board. Draw two children on each side. Ask students:

- **If one child gets off one side of the seesaw, is the seesaw balanced?**
- **Tell me what you would need to do to make the seesaw balanced again.**

Point to the original picture, and ask:

- **If one child gets on one side of the seesaw, is the seesaw balanced?**
- **Tell me what you would need to do to make the seesaw balanced.**

Write another equation on the board. Demonstrate with numbers how adding and/or subtracting changes equations. Write $2 + 2 = 4$. Then write $3 + 2 = 4$. Ask students:

- **Is the first number sentence correct?**
- **Is the second number sentence correct?**
- **Tell me what you need to do to make the second number sentence correct.**

Lesson Follow-Up

If students have difficulty completing this activity, have them complete Activity 27.

Activity 26

Independent **Objective**

Students identify information shown in a bar graph.

Review the term *bar graph* with students. Explain that graphs such as bar graphs help us see certain types of information. Many different types of information can be displayed on graphs. One type is change over time. On the board, draw a bar graph with five bars. Label the first bar *Monday,* the second *Tuesday,* and so forth through *Friday.* Label the horizontal lines of the graph *45/minute, 50/minute, 55/minute, 60/minute,* and *65/minute.* Say, "Here is a bar graph. This bar graph displays information about time. A student is preparing for a jump rope contest and practiced jumping the times shown per minute for a particular week. Look at the graph and answer the following questions:"

- **Show me/Tell me the number of times the student jumped rope in one minute on Monday.**
- **Show me/Tell me how many times the student jumped rope in one minute on Tuesday.**

Continue for all five days, and then ask students:

- **How did the information change over time?**

Lesson Follow-Up

If students have difficulty completing this activity, have them complete Activity 28.

Conceptual Development Activities

Activity 27

Developing **Objective**

Students compare two sets of objects (up to 25 in each set) using the terms more than, less than, same as, equal, and unequal.

Review the terms *equal/same as, more than,* and *less than.* Display two sets of equal objects, such as 25 pencils beside 25 pencils. Label one Side A and the other Side B. Ask a volunteer to count one side. Ask another volunteer to count the other side. Write the answers on the board. Then ask students:

- **Show me/Tell me if one side is the same as, or equal to, the other side.**
- **If I take a pencil away from Side A, are the sides equal?**
- **Does Side A have more than or less than Side B?**
- **If I add a pencil to Side A, are the sides equal?**
- **Does Side A have more than or less than Side B?**

Repeat this activity using other objects.

Lesson Follow-Up

If students have difficulty completing this activity, have them complete Activity 29.

Activity 28

Developing **Objective**

Students identify information on a pictograph.

Review the terms *object graph* and *pictograph.* Explain that graphs such as object graphs and pictographs help us see certain types of information. Many different types of information can be displayed on graphs. One type is change over time. On the board, draw a table/chart with four columns. Label horizontal lines vertically along the left edge of the chart. From bottom to top, label the lines *one inch, two inches, three inches,* and *four inches.* In each of the four columns, draw a beanstalk, left to right one inch high, two inches high, three inches high, and four inches high. Say, "Here is an object graph that shows how much the beanstalk plant grew each week for one month. Look at the graph and answer these questions."

- **How tall did the beanstalk grow in the first week?**
- **How tall did it grow in week two?**

Continue with all four weeks, and then ask students:

- **How did the beanstalk change over time?**

Lesson Follow-Up

If students have difficulty completing this activity, have them complete Activity 30.

Activity 29

Emerging **Objective**

Students group objects to create sets with the same quantity.

Review the terms *equal* and *same as*. Display objects (erasers, crayons, or building blocks, for example) on a desk or other surface area. Make sure there are equal numbers of all objects (six erasers, six crayons, and six blocks, for example). Continue by asking students:

- **Place all of the erasers together. Place all of the crayons together. Place all of the blocks together.**
- **Show me/Tell me how many erasers are in the set.**
- **Show me/Tell me how many crayons are in the set.**
- **Show me/Tell me how many blocks are in the set.**

Have students spend time looking at the sets and re-counting if necessary. Ask students:

- **Is the number of erasers the same as the number of crayons?**
- **Is the number of crayons the same as the number of blocks?**
- **Is the number of blocks the same as the number of erasers?**

Repeat this activity using other classroom objects.

Lesson Follow-Up

If students have difficulty with this activity, draw simple objects on the board (for example, 3 circles = 3 circles). Ask students if the number of shapes on the left of the equal sign is the same as the number of circles to the right of the equal sign.

Activity 30

Emerging **Objective**

Students review and recognize pictographs and object graphs.

Review the terms *object graph* and *pictograph*. Tell students that graphs help us see information using pictures. On the board, draw a three-column chart. Label the first column *Monday*, the second column *Wednesday*, and the third column *Friday*. Draw one apple in the first column, two apples in the second, and three apples in the third. Say, "Here is an object graph that shows how many apples students gave their teacher on three different days of the week. The apples represent information." Re-create the pictograph as an object graph on a desk or the floor. Explain that object graphs use objects to show information. On a set of index cards draw a bar graph, a line graph, an object graph, and a pictograph. Show students which is an object graph and which is a pictograph. Select 2 index cards, and ask students:

- **Show me/Tell me which graph is a pictograph.**
- **Show me/Tell me which graph is not an object graph.**

Lesson Follow-Up

If students have difficulty completing this activity, repeat the activity using various object graphs and pictographs.

Conceptual Development Activities

Activity 31

Independent Objective

Students use a number line to indicate position, using the terms before and after, of whole numbers up to 100.

Review the terms *number line* and *whole numbers* with students. Draw a number line from 1 to 10 on the board. Explain that numbers on this number line are all whole numbers. Point out that they are written in order from 1 to 10. Ask students:

- **Is the number 5 before or after the number 3?**
- **Is the number 6 before or after the number 9?**

If necessary, continue asking students "before and after" questions.

Draw another number line on the board, or display a premade number line from 0 to 100. Point to a specific number on the line, and have students identify which numbers come before. After several numbers, have students identify the number that comes after a particular number. On two index cards write the words *before* and *after*. Have a student say a number. Then hold up one of the cards, and have the students identify either the number before or after the stated number, based on the position listed on the card. For example, if a student says 65 and you are holding the *before* card, the student should say 64.

Lesson Follow-Up

If students have difficulty completing this activity, have them complete Activity 35.

Activity 32

Independent Objective

Students use tools to solve problems involving weight and length.

Review the terms *length, feet,* and *inches.* Display a ruler. Point out the 12 inches that make one foot. Explain that more than 12 inches would be more than one foot. Ask students:

- **If one ruler equals one foot, how many feet would be in two rulers?**
- **If 12 inches equals one foot, how many inches equal two feet?**

Remind students that rulers are used to measure things. Tell them you need to cut a picture to fit into a book. Hold up the ruler to the picture. Ask students:

- **How long should the page be for the picture to fit?**

Continue to ask similar length questions that require the students to measure.

Review the terms *weight* and *pounds*. Show students a scale, and explain how scales can be used to measure weight in pounds. Tell students you need to measure a box of books to mail. Place a box of books on the scale. Ask students:

- **How many pounds does the box of books weigh?**

Continue to ask similar weight questions that require the students to measure.

Lesson Follow-Up

If students have difficulty completing this activity, have them complete Activity 36.

Conceptual Development Activities

Activity 33

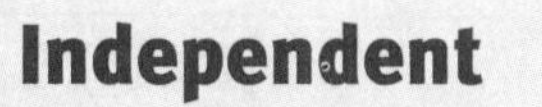

Independent **Objective**

Students identify time to the minute.

Review the terms *hour* and *minute*. Display a wall clock. Explain that the line markings between numbers indicate minutes. Tell students that if you counted all the lines, you would get 60. There are 60 minutes in one hour. Ask students to count with you the lines between 12:00 and 12:10. Review digital displays as well. Ask students:

- **Tell me how many lines we counted.**
- **Tell me how many minutes we counted.**
- **Tell me how many minutes there are between 12:00 and 1:00.**

Continue with various lengths of minutes. Then point to the clock and ask:

- **What time is it? How do you know?**
- **How would you show that time on a digital clock?**

Repeat the activity with a variety of times.

Lesson Follow-Up

If students have difficulty completing this activity, have them complete Activity 37.

Activity 34

Independent **Objective**

Students count to find the area of rectangles and squares on a grid.

Review the terms *square* and *rectangle* with students. Explain that shapes can be created on graph paper. The area of the shapes is determined by the number of squares inside the shape. Draw a grid on the board. Give each student graph paper with a rectangle and a square on it. Say, "In order to find the area of the shape on the page, we need to count the grid squares of the shape." Ask students:

- **Tell me how long the rectangle is in squares.**
- **How many squares is the rectangle in width?**
- **What is the area of the rectangle?**

Continue the activity using a square.

Lesson Follow-Up

If students have difficulty completing this activity, have them complete Activities 38 and 39.

Conceptual Development Activities

Activity 35

Developing Objective

Students identify the relative position of whole numbers (up to 10) on a number line using the terms before and after.

Draw a number line from 1 to 10 on the board. Explain that the numbers on this number line are all whole numbers. Point out that they are written in order from 1 to 10. Ask students:

- **Is the number 5 before or after the number 7?**
- **Is the number 5 before or after the number 4?**

If necessary, continue asking students "before and after" questions.

Display a number line labeled 0–10 in front of the students. Give each student a 0–5 number cube. Have the students announce either *before* or *after*. Then have the students roll the number cube. The students should then find that number on the number line, followed by either the number before or after depending on which they announced prior to rolling. Repeat this activity using a 5–10 number cube.

Lesson Follow-Up

If students have difficulty completing this activity, have them complete Activity 40.

Activity 36

Developing Objective

Students use tools to solve problems involving length and weight.

Review the terms *length, feet,* and *inches.* Display a ruler. Point out the 12 inches that make one foot. Have students count off the inches as you move your finger across the ruler. Remind students that rulers are used to measure things. Tell them they will measure the length of some objects. Hold up the ruler to a pencil, a pen, and an eraser. Say, "Here are the objects to measure." Then ask students:

- **How long is the pencil? The pen? The eraser?**
- **Could the pencil fit into a pencil box that is 4 inches long? 6 inches long? 8 inches long?**
- **My eraser is 3 inches long. Is the pen shorter than the eraser?**

Review the term *weight.* Show students a scale, and demonstrate how scales can be used to measure weight in pounds. Tell students they will weigh a box full of crayons. Place the crayon box on the scale and ask students:

- **How much does the box of crayons weigh?**
- **How do you know?**
- **What tool do you use to weigh?**

Lesson Follow-Up

If students have difficulty completing this activity, have them complete Activity 41.

Conceptual Development Activities

Activity 37

Developing **Objective**

Students identify time to the half-hour and hour.

Review the terms *hour* and *half hour.* Display a wall clock. Explain to students that one revolution of the big hand indicates one hour. A half revolution equals half an hour. Remind students that there are 60 minutes in an hour. Demonstrate how to tell time using an analog and digital clock. Show students various times on an analog clock face. Ask:

- **What time is it?**
- **If it is 4:30, show me/tell me where the long hand should point. Where should the short hand point?**

Repeat with a digital clock face. Next, create two sets of index cards showing various matching times. The first set of index cards should have times on a digital clock. The second set should have times on an analog clock. Have students match the digital time with the analog time. Students can also play a memory-style game in which they turn the cards over and look for matches.

Lesson Follow-Up

If students have difficulty completing this activity, have them complete Activity 42.

Activity 38

Developing **Objective**

Students informally find the perimeter of squares and rectangles.

Show students a rectangle made of cardboard. Have students trace the distance around the rectangle using their finger. Tell students to imagine that the rectangle is a garden, and you would like to know how much fence is needed to surround the garden. Lay interlocking cubes along the four sides of the rectangle. Say:

- **If I want to find out how much fence to buy to surround the garden, we add the side lengths together. Count with me the number of cubes on each side of this rectangle.**

Repeat this activity with other rectangles and squares. Next, show students a rectangle with two side lengths labeled 4 and two labeled 6. Demonstrate how to find the perimeter using addition. Say:

- **Another way to find the distance around an area is by adding. For this rectangle we would add 4 + 6 + 4 + 6, which equals 20.**

Show students other rectangles and squares, and have them add the side lengths together to find the perimeter.

Lesson Follow-Up

If students have difficulty completing this activity, have them complete Activity 43.

Conceptual Development Activities

Activity 39

Developing **Objective**

Students use physical models to compare the areas of two squares.

Display two squares made of base-ten cubes. One should be 4 by 4 and the other 6 by 6. Tell students that both are squares. Ask:

- **Show me/Tell me which square takes up more space.**

After students answer, have them count the number of cubes to confirm their answer. Say, "This square is made of 16 units, and this square is made of 36 units. The amount of space an object takes up is called area. To find area we can count the number of squares used to fill a space." Show students two rectangular figures made of base-ten cubes, one that is 2 by 6 and a second that is 3 by 4. Ask students which rectangle takes up more space. Have them confirm that both are equal by counting the cubes. Explain that some areas have different measures but take up the same area. Cut out various rectangles and squares from one-inch grid paper. Work with students to calculate the areas of the figures by counting the squares. Label the areas of the figures. Ask students:

- **Show me/Tell me which figure takes up the most space.**
- **Show me/Tell me which figure takes up the least space.**

Continue showing the various figures and asking students to compare the areas.

Lesson Follow-Up

If students have difficulty completing this activity, have them complete Activity 43.

Activity 40

Emerging **Objective**

Students use objects to count from 1 to 5.

Display five classroom objects (such as erasers, pencils, or building blocks) on a desk. Point to and count each object, clearly stating the numbers one through five. Repeat by having students count the number of objects with you. Say:

- **Point to each of the objects on the desk.**
- **Count each object on the desk.**
- **Tell me how many objects are on the desk.**

Repeat using 3 different objects. Set 3 objects in front of a student and ask:

- **Show me/Tell me how many objects there are.**

Count with students to confirm their answer. Repeat by adding or taking objects away and having students count them.

Lesson Follow-Up

If students have difficulty with this activity, draw five shapes, such as circles or squares, on the board. Count each of the shapes, and ask students to count with you.

Activity 41

Emerging **Objective**

Students identify the differences between objects, such as shape and size.

Draw a large circle and a small circle on the board. Beside it draw a large and small square. Explain that the circle and the square are different shapes. Point to the shapes, and trace them with your finger. Also explain how the same shapes can be different sizes. Show students real classroom objects that are these shapes, such as a can, globe, eraser, and a pencil box. Continue by asking students:

- **Show me/Tell me which objects look like this circle on the board.**
- **Show me/Tell me which objects look like the rectangle on the board.**

Display a hexagon and a triangle pattern block. Ask students:

- **A triangle has 3 sides. Show me/Tell me which shape is a triangle.**

Lesson Follow-Up

If students have difficulty with this activity, cut out large and small circles and rectangles from construction paper. Ask students to hold the shapes and trace them with their fingers. Ask students to point to the large and then the small shapes.

Activity 42

Emerging **Objective**

Students name the next activity in a daily schedule.

Display a daily schedule with classroom activities shown. If possible, use the students' actual schedule. Some activities could include reading time, sharing time, and writing time. Say, "Here is our schedule. It shows what we are going to do today. First we are going to read." Read a story to the students. After reading, ask:

- **Show me/Tell me what comes next.**

Students should indicate it is sharing time. Have students share something about what they did that morning or the day before. Continue through the list of activities. When you have completed them, show students pictures representing what you did. For example, a picture of a book could be used for reading time. Ask:

- **These pictures show what we did today. First we read a book. Show me/Tell me what we did next.**

Continue asking similar questions until the entire schedule has been covered.

Lesson Follow-Up

If students have difficulty with this activity, write the words *breakfast, lunch,* and *dinner* and/or *morning time* and *night time* on index cards. Have students place the activities in the order in which they occur in the day.

Conceptual Development Activities

Activity 43

Emerging **Objective**

Students identify the differences in area between two squares and two rectangles.

Display a small and a large square cut from one-inch grid paper. Say:

- **Show me/Tell me whether these squares are different.**
- **Show me/Tell me which square is smaller.**

Confirm answers by placing the smaller square on top of the larger square. Repeat this activity using 2 rectangles cut from one-inch grid paper. Ask students:

- **Show me/Tell me whether these rectangles are different.**
- **Show me/Tell me which rectangle is bigger.**

Confirm the answers. Repeat the activity. Include figures that are the same size to see if students understand that images can take up the same amount of space.

Lesson Follow-Up

If students have difficulty with this activity, cut out shapes in different sizes from construction paper. Allow students to hold the different shapes and sizes of shapes. As they are holding a particular shape, say "larger" or "smaller."

Activity 44

Independent **Objective**

Students identify multiples of 2, 5, and 10 for numbers to 100 through skip counting.

Display a number line from 0–10. Count by 2s with students starting at 0 and stopping at 10. Count by 2s again, but have students count with you as you mark the number line so they can see what counting by 2s looks like on a number line. Repeat by counting by 5s. Display a number line going from 0–30. Say:

- **We are going to count to 30 by 2s. Count with me.**

As students count, continue to show the skip counting on the number line. Repeat by skip counting by 5s and 10s. Ask:

- **Show me/Tell me how many jumps we made on the 0-30 number line when counting by 2s. 5s? 10s?**

Place an array of 100 objects on the desk, such as coins or counters. Say, "I'm going to group the coins by 2s so we can practice skip counting by 2s." Place 50 of the coins in groups of 2s. Then, as a group, count the coins on the desk. Ask students:

- **As we count, how many coins are added each time?**
- **How many total coins do we have when we count them all?**

Continue this process using different objects to practice counting by 2s, 5s, and 10s to 100.

Lesson Follow-Up

If students have difficulty completing this activity, have them complete Activity 50.

Conceptual Development Activities

Activity 45

Independent Objective

Students solve addition problems with three or more addends using the Associative Property.

Review the term *Associative Property* with students. Remind students that the associative property makes addition easier. Write the following equation on the board: $2 + 3 + 5 = 2 + 3 + 5$. Do the addition together with the class. Then write the following equation on the board:
$(2 + 3) + 5 = 2 + (3 + 5)$. Ask students:

- **How did this equation change from the previous equation?**
- **Tell me if you think we have the same answer as in the previous problem.**

As a class, solve the equation, adding the numbers in parentheses first. Ask students:

- **Do the numbers on the left side of the equal sign add up to the same as the numbers on the right side of the equal sign?**

Lesson Follow-Up

If students have difficulty completing this activity, have them complete Activity 51.

Activity 46

Independent Objective

Students use a number line to compare and order numbers to 100.

Display two number lines from 0 to 100 on the board or on the students' desks. Then, write several different numbers such as 12, 26, 75, and 92 on the board. Ask students:

- **A number line shows the position of numbers. I've placed some numbers on the number line. Show me/Tell me which number comes first on the number line.**

Ask students to place numbers such as 55, 63, and 98 on the number lines. When the numbers are placed, ask:

- **In what order do you place numbers on the number line?**
- **Is the number 55 before or after the number 63?**
- **Is the number 98 before or after the number 55?**

Continue practicing with numbers from 0 to 100 by having students select various index cards with numbers from 0–100 written on them. Have students compare the cards pulled and then order the cards on the number line.

Lesson Follow-Up

If students have difficulty completing this activity, have them complete Activity 52.

Conceptual Development Activities

Activity 47

Independent **Objective**

Students estimate and solve addition and subtraction problems in real world contexts and check for accuracy.

Review the term *estimating* with students. Remind them that when we estimate in addition and subtraction, we may not get the correct answer. When we estimate, we want to quickly find out a closer answer and then find the correct answer later. Say, "I went to the store to buy bread and milk. I knew that I had $6 with me. The bread cost $2 and the milk cost $3." Ask students:

- **Do I have enough money for both items?**
- **What is the estimated amount we're paying?**

Ask students to check their estimations by adding 2 and 3 and comparing the sum with 6. Repeat with similar questions using estimation with addition and subtraction problems.

Lesson Follow-Up

If students have difficulty completing this activity, have them complete Activity 53.

Activity 48

Independent **Objective**

Students solve addition and subtraction problems involving two-digit numbers and check for accuracy.

Remind students how to check for accuracy when doing addition and subtraction problems. Explain that they are going to solve both addition and subtraction problems and then use different methods to check their work. Say:

- **At the store I bought a shirt for $10. I had $25. How much money do I have left?**
- **Is this problem asking you to add or subtract?**

Write the following on the chalkboard: 25 – 10. Ask students:

- **What is the answer to this problem?**
- **Check your work with a calculator.**
- **Is your answer correct?**

Next ask:

- **I received 12 flowers for my birthday from my mother. I then received 12 more flowers from my brother. How many flowers do I have altogether?**
- **Is this problem asking you to add or subtract?**

Write the following on the board: 12 + 12. Then ask students:

- **Tell me the answer to this problem. Check your work using a calculator.**
- **Is your answer correct?**

Repeat using similar addition and subtraction problems.

Lesson Follow-Up

If students have difficulty completing this activity, have them complete Activity 53.

Conceptual Development Activities

Activity 49

Independent **Objective**

Students solve addition and subtraction problems with unknowns up to 18.

Explain to students that sometimes the answer is given in a problem but there is an unknown number in the equation. Say:

- **I'm going to give you 12 base-ten cubes. Then I'm going to give you some more. You now have a total of 18. Show me/Tell me how many additional cubes I gave you.**

Write the following on the board: $12 + x = 18$. Explain that in addition problems, it is necessary to use subtraction or the inverse of addition to find the missing number ($18 - 12 = 6$). Ask students:

- **To check our work, replace x with 6. Does $12 + 6 = 18$?**

Continue by saying,

- **You had a certain number of base-ten cubes. I'm going to take 8. Now you have 10. Show me/Tell me how many cubes you started with.**

Write the following: $x - 8 = 10$. Ask students,

- **To find the missing number, we add the total number of cubes. Show me/Tell me how many base-ten cubes you started with.** $10 + 8 = 18$
- **To check the answer, replace x with 18. Does $18 - 8 = 10$?**

Continue asking similar giving and taking questions. Be sure also to point out that in some subtracting situations subtraction is still used. For example:

- **You had 12 base-ten cubes. I'm going to take some. Now you have 5. Show me/Tell me how many cubes I took.**

This is written as $12 - x = 5$.

Lesson Follow-Up

If students have difficulty completing this activity, have them complete Activity 53.

Activity 50

Developing **Objective**

Students use skip-counting to count up to 30 by 5s.

Review the term *skip counting* with students. Display an array of objects on the desk, such as coins or small erasers. Place 30 objects so students can practice skip counting to 30. Say, "I'm going to group the coins by 5s so that we can practice skip counting by 5s." Place all of the coins in groups of 5s, or ask a volunteer to place the coins or objects in groups of 5s. Then count the coins on the desk as a class or ask a volunteer to come to the desk and skip count the coins. Ask students:

- **As we count, how many coins are added each time?**
- **Tell me how many coins we have when we count them all.**

Practice skip counting by 5s with other objects or by drawing groupings of 5 shapes on the board.

Lesson Follow-Up

If students have difficulty completing this activity, have them complete Activity 54.

Conceptual Development Activities

Activity 51

Developing **Objective**

Students solve addition problems using the Commutative Property to check for accuracy.

Review the term *Commutative Property* with students by displaying two groups of 12 counters. Set the first group in a set of 7 and a set of 5. Set the second group in a set of 5 and a group of 7. Ask:

- **Show me/Tell me whether these groups of counters are equal.**

Remind students that when solving addition problems, the order of the numbers does not matter. Write the following equations on the chalkboard: $5 + 7 = 12$; $7 + 5 = 12$. Ask students:

- **Tell me what the answer to the first equation is.**
- **What is the answer to the second equation?**
- **Are the answers to both equations the same?**

Say, "Here is another way to write this equation: $5 + 7 = 7 + 5$." Ask students:

- **What is the answer on both sides of the equal sign?**
- **Is the answer the same?**
- **Tell me if it matters what order the numbers are written. Are the answers the same?**

Continue practicing with other groups of counters and equations.

Lesson Follow-Up

If students have difficulty completing this activity, have them complete Activity 55.

Activity 52

Developing **Objective**

Students use objects, pictures, and a number line to compare and order whole number to 30.

Write a number line from 1 to 30 on the board. Fill in the numbers: 1, 15, and 30. Then write several numbers between 1 and 30 on the board. Say:

- **Place the numbers in the correct position on the number line.**

When the numbers are placed, ask:

- **In what order do you place numbers on the number line?**
- **Is the number 4 before or after the number 11?**
- **Is the number 25 before or after the number 12?**

Continue practicing ordering whole numbers to 30 using objects or pictures. Write the numbers 1 through 30 on index cards. Shuffle the cards. Ask volunteers to pick a card and place it on the board or on a desk. Ask a student to select another card and place it correctly before or after the first card. Continue until all cards have been used.

Lesson Follow-Up

If students have difficulty completing this activity, have them complete Activity 56.

Activity 53

Developing **Objective**

Students solve addition and related subtraction problems in real world contexts.

Tell students that they are going on a pretend shopping trip. Write on the board the items they will purchase. Beside the items write a dollar value for each one. Items and prices could include lunch box $8, bottled drink $1, and shirt $6. Also provide each student with $25 in play money. Say, "These are the items you can buy. Beside each item is how much it costs. You have $25 to spend. As you purchase each item, add up how much money you have spent. You must stop when you reach $25." Tell students to begin shopping, and ask them:

- **Show me/Tell me how much you spent on your first two items.**
- **Show me/Tell me how much money you had after buying your first two items.**
- **How many items were you able to purchase for your $25?**

Repeat the activity, but do not allow students to use money this time. Have students keep a running record of their work on a sheet of paper.

Lesson Follow-Up

If students have difficulty completing this activity, have them complete Activity 56.

Activity 54

Emerging **Objective**

Students count from 1 to 5 using objects and one-to-one correspondence.

Display 5 pencils. Count the pencils one at a time with the students and have them touch the pencils as you count them. Next, display two objects, such as counters, pencils, or markers. Say:

- **Here are two pencils. These are mine. I want to give you just as many pencils. Show me/Tell me when to stop giving you pencils.**

Students should indicate to stop when they also have two pencils. Continue by displaying various amounts from 1–5 of classroom objects and asking,

- **Here are three books. These are mine. I want to give you just as many books. Show me/Tell me when to stop giving you books.**

Lesson Follow-Up

If students have difficulty completing this activity, draw five simple shapes on the board. Count 1 to 5 for the students. Ask students to point to the objects as you say the number aloud. Then ask students to count with you.

Conceptual Development Activities

Activity 55

Emerging **Objective**

Students identify when items have been taken from or added to a set of up to 5 objects.

Display a grouping of five items, such as building blocks or pencils, on a desk or other surface. Say, "Here is a group of building blocks. We need to know how many blocks are on the desk." Count the number of items for the students. Then ask students to count aloud the number of items with you. Ask students:

- **Tell me how many building blocks we counted.**

Take away one of the blocks and put it behind the desk. Say, "I put one of the blocks away. Let's count to see how many blocks we have left." Count with students the number of blocks left on the desk. Ask students:

- **Tell me how many building blocks we have now.**

Continue taking away an item until you are down to one. Then say, "I'm going to put two blocks back." Count the number of blocks with students. Ask students:

- **How many blocks do we have now?**

Continue practicing with various items.

Lesson Follow-Up

If students have difficulty with this activity, hand individual students up to 5 objects at a time. Add and take away objects and count them as you go.

Activity 56

Emerging **Objective**

Students solve problems using the terms more, less and same.

Review the terms *more, less,* and *same* with students. Display 3 blocks and 6 blocks. Say:

- **I have 6 blocks. You have 3 blocks. Show me/Tell me who has more blocks.**

Display 4 marbles and 10 marbles. Ask:

- **I have 4 marbles. You have 10 marbles. Show me/Tell me who has more marbles.**

Display 5 cups and 5 cups. Ask:

- **I have 5 cups. You have 5 cups. Show me/Tell me who has more cups. We have the same amount.**

Continue displaying objects and asking similar questions using the terms *more, less,* and *same.*

Lesson Follow-Up

If students have difficulty with this activity, have them hold items of unequal amounts and weights. Ask them to indicate which item has more by feeling how heavy it is. Point out that the heavier items have more quantities and the lighter items have less.

Conceptual Development Activities

Activity 57

Independent **Objective**

Students sort and count data before displaying data on a pictograph and bar graph.

Review the terms *data, pictograph,* and *bar graph* with students. Remind students that information can be displayed in graphs to make it easier to see and understand. Cut out pictures of three different school activities (for example, books for reading, numbers for math, and a pencil for writing). Ask:

- **Select a picture of your favorite school activity.**

Discuss what each student chose. Say,

- **Three people like math. Four people like reading. Two people like writing.**

On the board display the frame for a pictograph. Explain that a pictograph is a graph using pictures. The pictures represent data. Affix the students' pictures to the board. Create the frame for a bar graph. Ask quantity questions such as

- **Show me/Tell me how many people responded that math is their favorite activity.**

As students answer the questions, create bars showing the data. Repeat this activity on a different topic such as favorite ice cream flavor or favorite movie.

Lesson Follow-Up

If students have difficulty completing this activity, have them complete Activity 59.

Activity 58

Independent **Objective**

Students describe the meaning of data in a pictograph and bar graph.

Review the terms *data, pictograph,* and *bar graph* with students. Remind students that information can be displayed in graphs to make it easier to see and understand. Color index cards one of three different colors (for example, red, blue, and green). Ask:

- **Select an index card showing your favorite color.**

Discuss what each student took. Say:

- **Three people like red. Four people like green. Two people like blue.**

On the board display the frame for a pictograph. Explain that a pictograph is a graph using pictures. The pictures represent data. Affix the students' index cards to the board. Create the frame for a bar graph. Ask quantity questions such as

- **Show me/Tell me how many people responded that green is their favorite color.**

As students answer the questions, create bars showing the data. Ask students:

- **Show me/Tell me which color people like the best.**
- **Show me/Tell me which color people like the least.**

Repeat this activity using various topics.

Lesson Follow-Up

If students have difficulty completing this activity, have them complete Activity 60.

Conceptual Development Activities

Activity 59

Developing **Objective**

Students sort objects into categories and display the data in a pictograph.

Review the terms *data, pictograph,* and *object graph*. Remind students that information can be displayed in graphs to make it easier to understand. Display two objects such as fruit (bananas and apples) or markers (orange and blue). Ask students to select their favorite fruit or a marker of their favorite color. Then arrange the objects in rows. Explain that by doing this they are creating an object graph. When the objects are sorted and arranged, ask:

- **Show me/Tell me how many people like the color orange.**
- **Show me/Tell me how many people like apples.**

Create a pictograph on the board representing the same data shown on the object graph. Explain that a pictograph uses pictures instead of real objects, but the results are the same. Ask students:

- **Count the number of apples. Is the number of apples the same in the pictograph as in the object graph?**
- **Show me/Tell me how many people like the color blue.**

Repeat this activity using various topics.

Lesson Follow-Up

If students have difficulty completing this activity, have them complete Activity 61.

Activity 60

Developing **Objective**

Students read two-category pictograph and object graph and identify the meaning of the data.

Review the terms *data, pictograph,* and *object graph*. Remind students that information can be displayed in graphs to make it easier to understand. Display two objects such as fruit (bananas and apples) or markers (orange and blue). Ask students to select their favorite fruit or marker of their favorite color. Then arrange the objects in rows. Explain that by doing this they are creating an object graph. When the objects are sorted and arranged, ask:

- **Show me/Tell me which color people like the most. Which do they like the least?**
- **Show me/Tell me which fruit do people like most. Which do they like the least?**

Create a pictograph on the board representing the same data shown on the object graph. Explain that a pictograph uses pictures instead of real objects, but the results are the same. Ask students:

- **Count the number of apples. Is the number of apples the same in the pictograph as in the object graph?**
- **Show me/Tell me how many more people like blue than orange.**

Repeat this activity using various topics.

Lesson Follow-Up

If students have difficulty completing this activity, have them complete Activity 61.

Activity 61

Emerging **Objective**

Students count up to 5 using pictures from a pictograph.

Remind students that information can be displayed in graphs to make it easier to understand. Tell students that a survey was taken to determine students' favorite foods for lunch. On a desk, show three jars of peanut butter to represent peanut butter sandwiches, five colored triangles resembling pizza to represent pizza slices, and two boxes of macaroni and cheese to represent macaroni and cheese. Ask students:

- **Show me/Tell me the number of jars of peanut butter shown.**
- **Show me/Tell me the number of pizza slices shown.**
- **Show me/Tell me the number of macaroni and cheese boxes.**

On the board draw a three-column chart with the headings *peanut butter sandwich, pizza,* and *macaroni and cheese.* Under *peanut butter sandwich* draw three sandwiches, under *pizza* draw five slices of pizza, and under *macaroni and cheese* draw two pieces of macaroni. Explain that each item represents one student. Ask:

- **Show me/Tell me the number of peanut butter sandwiches shown.**
- **Show me/Tell me the number of pizza slices shown.**
- **Show me/Tell me the number of macaroni and cheese boxes.**

Lesson Follow-Up

If students have difficulty with this activity, take a survey of the class and record their favorite lunchtime foods on a pictograph on the board. Have students point to their favorite choice.

6

Conceptual Development Activities

Activity 1

Independent Objective

Students express and represent fractions, including halves, fourths, thirds, and eighths.

Using two-colored counters, place four counters in front of the students. Have three red and one yellow. Demonstrate how this shows one fourth and how to write $\frac{1}{4}$. Continue with this model to show halves, thirds, and eighths. Give students two counters each: one red, one yellow.

- **Show/Tell me what fraction this is.**
- **Write the fraction.**

Continue asking students to identify thirds, fourths, and eighths. Next, give students eight counters each.

- **Show me one half.**
- **Write the fraction.**

Continue asking students to model and write different fractions of thirds, fourths, and eighths.

Lesson Follow-Up

If students have difficulty completing this activity, have them complete Activity 5.

Activity 2

Independent Objective

Students identify multiplication as repeated addition of equal groups using physical models.

Give each student six counters. Demonstrate how to divide the counters into three groups of two. Say, **Here are three groups of two.**

- **Tell me how many counters you have.**

Draw three circles on the board and put two Xs in each circle. Have students copy the pictures on the whiteboard. Next, show how to write $2 + 2 + 2 = 6$. Practice this procedure with different multiplication facts until students demonstrate mastery. Then have students place counters into groups following verbal prompts:

- **Show me three groups of four.**
- **Draw a picture of three groups of four.**
- **Tell me an addition problem for three groups of four.**

Follow this process with other multiplication facts. Then introduce the idea that "groups of" is the same as "times." Repeat the above steps with multiplication facts. For example, say, **Show me 3 × 4.** Have students solve a real-world problem using multiplication.

- **How many legs do three horses have?**

Lesson Follow-Up

If students have difficulty completing this activity, have them complete Activity 6.

Conceptual Development Activities

Activity 3

Independent **Objective**

Students identify division as repeated subtraction of equal groups using physical models.

Give each student eight linking cubes of the same color, and have them make a train.

- **How many cubes do you have altogether?**
- **How could you give two people the same number of cubes?**

Demonstrate how you can break the train into two groups. Put the trains back together.

- **How many cubes do you have in your train?**
- **Show me how to break off a group of two.**
- **How many cubes do you have left in your train?**

Show the students how they have just demonstrated $8 - 2 = 6$. Direct students to continue to break off groups of two, and show them the subtraction problem for each step. When the train is completely broken into groups of two, ask:

- **How many groups do you have?**

Write $8 \div 2 = 4$, and show students how they have just divided. Repeat the procedure for other simple division facts.

Lesson Follow-Up

If students have difficulty completing this activity, have them complete Activity 7.

Activity 4

Independent **Objective**

Students solve real-world problems involving fractions, including halves, fourths, thirds, and eighths.

Show students a set of measuring cups including 1 cup, $\frac{1}{2}$ cup, $\frac{1}{3}$ cup, and $\frac{1}{4}$ cup. Let them explore the measuring cups by putting them in order from largest to smallest.

- **Show me one cup. Show me one-half cup.**
- **Show/Tell me which is bigger.**

Continue making comparisons to make students comfortable with the measurements.

- **How many half cups will fit into one cup?**
- **Why do you think that?**

After the students all make a prediction, fill the one-cup measuring cup using the one-half-cup measuring cup. You can use water, sand, or rice to do this. Discuss how two halves make a whole. Follow this procedure for $\frac{1}{3}$ cup and $\frac{1}{4}$ cup. To check for understanding, ask students how many $\frac{1}{8}$ cups would fit into one cup.

Lesson Follow-Up

If students have difficulty completing this activity, have them complete Activity 8.

Activity 5

Developing **Objective**

Students express, represent, and use fractions, including halves, fourths, and thirds, as parts of a whole and as parts of a set.

Give each student a set of fraction circles with the following pieces: one whole and two halves. Let them explore relationships for a few minutes. Then have them place the two halves on top of the whole.

- **How many halves does it take to make a whole?**

Next have students take one half away.

- **How many halves do I have left?**
- **Here is one half. Say *one half*.**

Now demonstrate how to write $\frac{1}{2}$. Explain how the fraction represents part over whole. Continue with this model using thirds and fourths. At this point, ask the students to do the following:

- **Show me three fourths.**
- **Write $\frac{3}{4}$.**

Continue with other fractions using halves, thirds, and fourths. Have students make their own fractions. Prompt them:

- **Tell me what fraction this is.**
- **Write the fraction.**

When students are comfortable using the fraction circles, repeat the activity using sets of four counters to demonstrate fractions as parts of a set.

Lesson Follow-Up

If students have difficulty completing this activity, then have them complete Activity 9.

Activity 6

Developing **Objective**

Students combine equal sets with quantities to 30.

Give each student twelve color tiles. (Each student should have only one color.) Have students arrange the tiles into squares and rectangles on one-inch grid paper. Model making a two-by-six array. Say, **Here are two groups of six.**

- **Show me two groups of six.**

Have students color the squares they have covered. Explain how two groups of six is the same as $6 + 6 = 12$. If they have trouble seeing this, make one row a different color. Continue this process to show other multiplication problems. Next, make a 3-by-4 array of color tiles. Write on the board $2 + 2 + 2 = 6$; $4 + 4 + 4 = 12$; $3 + 3 + 3 = 9$

- **Show/Tell me which is the correct addition problem.**

Follow this model for other simple multiplication problems. Follow up with a real-world example, such as

- **How many shoes are there in this room?**

Have students write an addition problem to answer the question.

Lesson Follow-Up

If students have difficulty completing this activity, have them complete Activity 9.

Conceptual Development Activities

Activity 7

Developing **Objective**

Students separate (divide) quantities using objects or pictures.

Give each student twelve counters and a piece of paper divided into fourths. Show students how to evenly divide the counters onto each section of the paper. Then say, **Now you show me how to divide the twelve counters evenly.**

- **How many counters are there in each square?**

Direct students to draw an X in place of each counter. Ask:

- **How many Xs are there altogether?**
- **How many Xs are there in each square?**

Continue this procedure with different division facts. As students demonstrate mastery, have them solve the problem without the manipulatives. Say:

- **Show me how to divide twelve into four equal groups with a picture.**
- **How many Xs are in each group?**

Use this procedure for other division facts. Follow this up with a real-world application:

- **How would you divide twelve snacks equally among four children?**

Lesson Follow-Up

If students have difficulty completing this activity, have them complete Activity 9.

Activity 8

Developing **Objective**

Students solve real-world problems involving fractions, including halves, fourths, and thirds, using objects.

If possible, use individual-size pizzas. If not, make some paper pizzas. Say to the students, **Here is one whole pizza.**

Discuss how you could divide the pizza between two people. Point out different ways to cut the pizza into two pieces. Ask students if it would be fair if one person got more pizza than the other.

- **Show me how I can divide this pizza into halves.**
- **Is each piece the same size?**
- **Show/Tell me how much of the whole pizza each person gets.**

Follow the above procedure to divide the pizzas into thirds and fourths. Leave one pizza whole. Have all the pieces of pizza in front of the students.

- **If you could only have one piece, which one would you choose? Why?**
- **What could we do to make sure everyone gets a whole pizza?**

Guide students to realize that two halves make one whole.

Lesson Follow-Up

If students have difficulty completing this activity, have them complete Activity 10.

Conceptual Development Activities

Activity 9

Emerging Objective

Students recognize half of sets of objects up to 4.

Give each student a piece of paper folded in half. Explain that this is one piece of paper that is divided into two halves. Have students cut along the fold or cut along the fold for them. Hold up one half, and say, **Here is one half of the paper.** Have students hold up one-half of the paper.

Have students place both halves in front of them with a space between them. Give each student two counters. Show them how to place one counter on each piece. Practice this several times. Explain that they have divided the two counters equally and that one half of two is one.

- **Show me one half.**

Repeat the process with four counters.

Lesson Follow-Up

If students have difficulty completing this activity, break pretzel sticks in half in front of the students for them to share. Follow this with creating situations in which two students have to equally share two or four objects. Reinforce the concept of *half* whenever appropriate.

Activity 10

Emerging Objective

Students solve simple problems involving joining and separating.

Give each student two linking cubes trains. One train is two cubes of one color. The second train is two cubes of another color.

- **How many cubes do you have of the first color?**
- **How many cubes do you have of the second color?**

Join the two trains together.

- **How many cubes do you have altogether?**

Separate the two groups and repeat the above questions. Write $2 + 2 = 4$ and show students how to use linking cubes to represent the problem. Say:

- **Show me $2 + 2 = 4$.**

Create a different train of four cubes, 2 of one color and two of another color.

- **How many cubes do you have altogether?**

Separate the train into two trains.

- **How many cubes do you have of the first color?**
- **How many cubes do you have of the second color?**

Follow the same procedure for different sets. Find real-world situations to reinforce this concept. For example, say:

- **How many wheels are on two cars?**

Lesson Follow-Up

If students have difficulty completing this activity, use familiar objects. For example, put 2 markers and 2 crayons in front of students. Ask how many of each and then how many things you can color with.

Conceptual Development Activities

Activity 11

Independent **Objective**

Students identify the meaning of common ratios using money.

Show students a penny, a nickel, a dime, and a quarter. Hold up a penny and a nickel again and say, **Here is a penny. It is worth one cent. Here is a nickel. It is worth five cents.**

Explain how you can use five pennies to equal one nickel. Next give each student a handful of pennies. Demonstrate how to trade five pennies for one nickel. Tell students:

- **Show/Tell me how you can trade your pennies for nickels.**

Students should then trade all the pennies they can for nickels. Check for understanding by observing what they do when they don't have five pennies left at the end to trade. Continue this process with the dime and the quarter. This activity can be extended by comparing coins to a dollar. For example: There are four quarters in a dollar, ten dimes in a dollar, twenty nickels in a dollar, and one hundred pennies.

Lesson Follow-Up

If students have difficulty completing this activity, have them complete Activity 13.

Activity 12

Independent **Objective**

Students identify rate including a measure of speed and a measure of cost.

Set up a class store with objects the students are familiar with. Make everything the same price such as one penny. Give each student 5 pennies. Say, **One pencil is a penny.**

- **How much are two pencils?**

Show them this by putting a penny with each pencil so they can see the relationship. Then have them count the pennies. Continue this model using different quantities. When they demonstrate mastery of this concept, let them go shopping. Ask questions as they shop:

- **How much does one item cost?**
- **How many items are you buying?**
- **How much do they cost together?**

Follow this by having each item cost two pennies. Follow the procedure above to implement the price change.

Once students understand the concept of rate, give them other problems to solve such as:

- **Apples cost $1 per pound at the grocery store. How much do 2 pounds cost?**
- **One gallon of gas costs $2. How much do 2 gallons cost? How much do 3 gallons cost?**

Next, help students understand measures of speed. Say, **In one hour I can drive 20 miles.**

- **How far can I drive in 2 hours?**

If necessary, give students counters to model the number of miles driven in each hour.

Lesson Follow-Up

If students have difficulty completing this activity, have them complete Activity 14.

Conceptual Development Activities

Activity 13

Developing **Objective**

Students recognize the meaning of a simple ratio, such as *2 to 1*.

Place one block in front of each student, saying:

- **Here is a block for student 1.**
- **Here is a block for student 2.**

When each student has a block, ask:

- **How many blocks does each person have?**

Next give each student one more block following the model above:

- **How many blocks does each person have now?**

Follow this procedure until each student has four blocks. Then give one student enough blocks for each student to have one block. Say, **Give each student one block.**

Repeat this step so each student has a chance to distribute the blocks. Follow the same process for two, three, and four blocks. Now show them pictures, toys, or actual everyday objects. For example, show bicycles, tricycles, and cars.

- **How many wheels does each bicycle have?**
- **How many wheels does each tricycle have?**
- **How many wheels does each car have?**

Lesson Follow-Up

If students have difficulty completing this activity, have them complete Activity 15.

Activity 14

Developing **Objective**

Students identify rate, including how dast something happens.

Find a physical activity that the students are able to do, such as bouncing a ball, jumping rope, hopping, or jumping jacks. Different students can do different activities.

- **How many times do you think you can bounce a ball in a minute?**

Then time them for one minute doing their chosen activity. Compare how many times they thought they could bounce the ball with what they actually did.

- **How many times did you bounce the ball in 1 minute?**
- **Can you bounce the ball faster?**

Time them again. Compare the number of bounces.

- **How many times did you bounce the ball in 1 minute?**
- **Which time did you go faster?**
- **How do you know you were faster?**

Should a student not be able to improve his or her speed, let the student practice the activity for several days and try again. Also, if a student is not able to sustain the activity for a full minute, cut down the time.

Lesson Follow-Up

If students have difficulty completing this activity, have them complete Activity 16.

Conceptual Development Activities

Activity 15

Emerging **Objective**

Students recognize differences in quantity in two sets of objects.

Show students two trains of linking cubes, one with 2 cubes and one with 4 cubes. Say, **Here is one train. Count how many cubes are in it.**

- **Show/Tell me how many cubes there are.**

Repeat the process with the other train. Next lay the trains side-by-side and say:

- **Show/Tell me which one is longer.**

Use the number line to show them that four is bigger than two. If students have never used a number line, demonstrate how to identify numbers on the number line. Show how the bottom two linking cubes in the train with four have a "partner" in the train of two.

- **Show/Tell me how many cubes don't have a partner.**

Repeat this step with each student. Continue this activity with different-sized trains up to six. The students may need prompts such as:

- **Does this cube have a partner?**

If appropriate, show students how to count from two to four on the number line.

Lesson Follow-Up

If students have difficulty completing this activity, use everyday objects such as pencils. Give one student more pencils than the other. Ask who has more. Ask how many more pencils the student with less would need to have the same amount as the student with more.

Activity 16

Emerging **Objective**

Students combine equal sets with quantities to 30.

Place in front of the students an object that can move, such as a toy car or a ball. Move the car very slowly, and say:

- **Here is slow.**

Now move the car very fast, and say:

- **Here is fast.**

Continue alternating between fast and slow until students appear to understand. Now move the car either fast or slow, and say:

- **Tell me whether the car is fast or slow.**

When they show mastery of this concept, show them how to walk slowly and how to walk fast using the model above. When the students understand the difference between walking fast and slow, say:

- **Show me fast.**
- **Show me slow.**

Then have one student walk either fast or slow and have the other students say whether it is fast or slow. Continue so that each student gets a turn. Students may perform other movements at different speeds, such as waving their hands or talking.

Lesson Follow-Up

If students have difficulty completing this activity, have one student walk while another student runs. Discuss how the runner is fast and the walker is slow. Reinforce the concept of *fast/slow* whenever appropriate.

Conceptual Development Activities

Activity 17

Independent **Objective**

Students write and solve number sentences (equations) involving addition and subtraction with two-digit numbers.

Using a sales ad from a grocery or dollar store, select three items to "buy" that are under $1.00. Write down the prices, leaving a space between amounts (34¢ 59¢ 67¢). Say, **To find out how much money I need to buy these three things, I will write a number sentence. I need to add all three prices together.**

Insert addition signs and an equal sign to complete the number sentence. Demonstrate using a calculator to solve the number sentence. Write the answer. Together, choose three new items to buy. Say:

- **Let's write a number sentence using the prices of the things we want to buy.**
- **Now let's solve the number sentence on the calculator.**
- **Let's write the answer.**

Repeat several times using new prices; then have students choose their own items.

- **Show me your number sentence.**
- **Show me how you find the answer with a calculator.**
- **Tell me the answer.**

Once students understand the concept of addition, modify the activity to model subtraction. Use model money to "pay" for their purchases, one item at a time.

Lesson Follow-Up

If students have difficulty completing this activity, have them complete Activity 22.

Activity 18

Independent **Objective**

Students use models and diagrams to solve problems with inequalities, including the > and < signs.

Use two different sets of items (such as 3 toy fish and 5 toy birds). Place the sets in front of the students. Show that there are three fish and five birds. Write the numbers 3 and 5 on the board, leaving a space for the < or > sign. Using a number line, ring the number 3, and say "three fish." Ring the number 5 and say "five birds." Show on the number line that 3 is less than 5, and write the < sign to complete the inequality. Continue in this way, varying the number of items in each set. Then give students sets of items, a number line, and space on the board.

- **Show/Tell me how many fish.**

Write the number on the board.

- **Show/Tell me how many birds.**

Write the number on the board.

- **Show me the numbers on the number line.**
- **Show/Tell me if we use < or > in the number sentence.**

Once students have mastered the concept using models, repeat the activity using diagrams or picture cards showing different quantities.

Lesson Follow-Up

If students have difficulty completing this activity, have them complete Activity 23.

Conceptual Development Activities

Activity 19

Independent Objective

Students identify function rules with addition and subtraction of one digit numbers.

Distribute linking chains to the students. Have them make chains of 2, 4, 6, and 8. Demonstrate how to line up the chains one above the other from the least to the most. Check to make sure that the students have their chains in the correct order and that they are lined up on the left. As a group, have the students point to each chain and say 2, 4, 6, 8.

- **Tell me, how many more links are in the chain that has 4 links than are in the chain that has 2 links?**

Demonstrate counting the difference between the two chains. Repeat as needed. Follow this process to show the differences between the chains. When the students demonstrate understanding of this concept, introduce the function rule of +2. Repeat the activity with different function rules.

Lesson Follow-Up

If students have difficulty completing this activity, have them complete Activity 24.

Activity 20

Independent Objective

Students use the Commutative and Associative Properties of addition.

Start with several red crayons and blue crayons. Take 4 blue crayons and 3 red crayons. Say, **I have four blue crayons and three red crayons. I have seven crayons altogether.** Now take 3 blue crayons and 4 red crayons. Say, **Now I have three blue crayons and four red crayons. I have seven crayons altogether.**

Write the two addition problems. Say, **Four plus three and three plus four equal seven.**

Next, give the students several red and blue crayons. Model each step as you say it.

- **Get two red crayons and three blue crayons.**
- **Show/Tell me, how many crayons do you have altogether?**
- **Now, get three red crayons and two blue crayons.**
- **Show/Tell me, how many crayons do you have altogether?**
- **Show/Tell me two ways to write *two and three are five.***

Repeat this process as needed. Have students complete similar problems independently. When students have mastered this skill, introduce Associative Properties. Use the same method as above, but add a third color of crayon.

Lesson Follow-Up

If students have difficulty completing this activity, have them complete Activity 26.

Conceptual Development Activities

Activity 21

Independent Objective

Students solve addition and subtraction number sentences using physical models, diagrams, and tables.

Determine how many girls in class prefer oranges, how many girls prefer apples, how many boys prefer oranges, and how many boys prefer apples. Have available real or play fruit, and group the fruit into sets based on students' responses. Say, **I want to know how many more boys than girls like oranges best.**

Demonstrate how to find the information and set up the number sentence. Solve the equation, and repeat the question in statement form with the answer. Say, **Three boys like oranges. One girl likes oranges. Two more boys than girls like oranges best.**

Repeat the process with an "altogether" question, having the students assist in finding the information and setting up the number sentence.

Next, work with students to set up a diagram or chart that reflects the same survey information. Again, ask students questions about the information, and help them set up addition and subtraction number sentences.

- **Find the information on the chart.**
- **Show me the number sentence.**
- **Tell me the answer.**

Lesson Follow-Up

If students have difficulty completing this activity, have them complete Activity 25.

Activity 22

Developing Objective

Students write and solve number sentences corresponding to real-world problem situations involving addition and subtraction with one-digit numbers.

Give two students different amounts of similar objects. For example, give Ann 3 cat figures and Bob 5 dog figures. Have the class help the two students count how many "pets" each one has. Say, **Ann has 3 cats.** Write 3 on the board. Say, **Bob has 5 dogs.** Write 5 on the board. Say, **Let's find out how many pets Ann and Bob have altogether.**

Write an addition sign and equal sign to complete the number sentence. Read it aloud. Have students count the cats and dogs aloud as Ann and Bob put their "pets" in a "pen" one at a time. Write the answer, and read the number sentence.

Repeat this activity using different "pets." Have one student give some "pets" to another student and set up a subtraction number sentence. Increase students' independence by pairing them with a partner to write and solve different number sentences.

Lesson Follow-Up

If students have difficulty completing this activity, have them complete Activity 27.

Conceptual Development Activities

Activity 23

Developing Objective

Students solve problems with inequalities, including the terms *more than* and *less than.*

Give each student a number line. Point to and say each number in order on the number line. Have the boys in the class raise their hands. Count aloud how many boys there are. Write down the number of boys on the board (for example, 12). Repeat with the girls (for example, 5). While demonstrating, say:

- **Find the number 12.**
- **Put your finger on it.**
- **Now, find the number 5.**
- **Put your finger on it.**
- **12 is *more than* 5.**
- **There are more boys than girls.**

Repeat this activity comparing a variety of different things of interest to the students. Include both physical models and diagrams or picture cards in your examples. Have the students take turns choosing what to compare.

Lesson Follow-Up

If students have difficulty completing this activity, have them complete Activity 27.

Activity 24

Developing Objective

Students identify function rules of *1 more* and *1 less.*

Give each student a number line. Have the students repeat your action as you show and say:

- **Here is the number 4.**
- **Show me the number 4.**
- **One more than 4 is 5.**
- **Show me the number 5.**
- **Put your finger back on the 4.**
- **One less then 4 is 3.**
- **Show me the number 3.**

Repeat this activity starting with different numbers. Gradually reduce the amount of support given until students are independent.

Lesson Follow-Up

If students have difficulty completing this activity, have them complete Activity 28.

Conceptual Development Activities

Activity 25

Developing Objective

Students use information from models, diagrams, tables, or pictographs to solve number sentences involving addition and subtraction with one-digit numbers.

Display a table or pictograph that represents, for example, the number of students who have a dog, and the number of students who have a cat. Review the information with the students. Ask the question:

- **How many more students have dogs than cats?**
- **What numbers do you know?**
- **What number do you need to find out?**
- **Do you need to add or subtract?**

Show students how to find the information needed and how to set up the number sentence. Have students calculate the answer using whatever method is necessary. Repeat the activity with various types of information. Students may compare sets of classroom objects, compare the number of sides on different geometric shapes, and so on.

Lesson Follow-Up

If students have difficulty completing this activity, have them complete Activity 28 (for MA.6.A.3.4), or Activities 27 and 30 (for MA.6.A.3.6).

Activity 26

Developing Objective

Students use the Commutative Property of addition.

Start with several red and blue bears or counters of different colors. Say, **I will get three red bears and four blue bears. I have seven bears altogether.**
Now, get four red bears and three blue bears. Say, **Now I have four red bears and three blue bears. I have seven bears altogether.**

Write the two addition problems. Say:

- **Three plus four and four plus three equal seven.**

Give students several red and blue bears. Model each step as you say:

- **Get two red bears and three blue bears.**
- **Tell me, how many bears do you have altogether?**
- **Now, get three red bears and two blue bears.**
- **Tell me, how many bears do you have altogether?**
- **Show me two ways to write *two and three is five.***

Repeat this process as needed. When students are comfortable solving problems with physical models, have them solve problems using number cards or other visual models. Then, have students complete similar problems independently.

Lesson Follow-Up

If students have difficulty completing this activity, have them complete Activity 29.

Conceptual Development Activities

Activity 27

Emerging **Objective**

Students solve simple problems using language such as *more, less, same,* and *none.*

Use two sets of very different objects, such as two spoons and three pencils. Place the sets in front of the students. Counting aloud, show that there are two spoons and three pencils. Say, **There are more pencils than spoons.** Continue in this way, demonstrating *more, less, same,* and *none*. Use one-to-one correspondence to help students compare the sets. It's a good idea to change objects used as well. Give students sets of items. Say, **Here are three spoons. Here are two pencils.**

- **Show/Tell me which is more, spoons or pencils.**

Remember to point to the objects as you say their names.

Repeat the activity, varying the type and quantity of items.

Lesson Follow-Up

If students have difficulty completing this activity, work with *more, less, same, and none* on several occasions using classroom objects (more books, more pencils, or more chairs). As often as appropriate, reinforce the concepts with all students.

Activity 28

Emerging **Objective**

Students combine equal sets with quantities to 30.

Use seven matching objects, such as erasers. Set out three erasers. Say, **Let's count the erasers. One, two, three. Let's add one more.**

- **Now how many do we have?**

Say, **Let's count. One, two, three, four.**

Repeat the process using different starting amounts up to six, always adding one more.

Set out 2 groups of objects such as a group of 2 erasers and a group of 3 erasers.

- **Show me three erasers.**

Say, **Add one more.**

- **Tell me how many erasers you have now.**

Lesson Follow-Up

If students have difficulty completing this activity, use objects that appeal strongly to the student. If appropriate, try a food item such as crackers or raisins. Set out two items on a plate, and tell students you gave them two raisins. Ask if they would like one more. Put one more on the plate and count 1, 2, 3. Let the students eat the raisins and restate that you gave them two raisins, then one more; therefore, they ate three raisins. Reinforce the concept whenever appropriate.

Activity 29

Emerging **Objective**

Students determine if the quantity in two sets of objects to six is the same or different.

Set out two sets of three objects that are the same. Arrange both sets in the same manner. Show and tell the students that the two sets are the same. Set out one set of two objects and one set of four objects. Show and tell the students that the sets are different. Give the students two sets of four objects. Demonstrate, and have the students arrange the sets in the same way.

- **Tell me if the sets are the same or different.**
- **How do you know?**

Repeat the process as many times as needed. If necessary, use one-to-one correspondence to help students compare the sets.

Finally, ask students to make two sets that are the same and two sets that are different.

Lesson Follow-Up

If students have difficulty completing this activity, help students compare two different items such as one shoe and one pencil. Say, "These are different." Then compare two of the same items, such as two pencils. Say, "These are the same." Reinforce the concepts of *same* and *different* whenever possible.

Activity 30

Emerging **Objective**

Students determine if the quantity in two sets of objects to 6 is the same or different.

Display a set of 3 pencils and a set of 5 pencils. Count the pencils in each set, and say, **This set has 3 pencils. This set has 5 pencils. They are different.**

Display two sets of 4 pencils. Count the pencils in each set, and say, **This set has 4 pencils and this set has 4 pencils. They are the same.**

Repeat the process but ask the student:

- **Tell me, are they the same or different?**

If necessary, help students use one-to-one correspondence to compare the sets. Repeat the activity, using different quantities to 6.

Lesson Follow-Up

If students have difficulty completing this activity, gather a variety of different-colored socks. Talk about *same* and *different.* Have the student make sets of two socks, indicating whether they are the same or different.

Conceptual Development Activities

Activity 31

Independent **Objective**

Students compare area and circumferences of different circles.

Prepare several different-sized circles cut from tag board. Place them in front of the students. Say, **Here are four circles.**

- **Which one is the biggest?**
- **Which one is the smallest?**

Ask students to put them in order from biggest to smallest.

Have the students place the circles on top of each other to see if they were correct. After each student has had a chance to arrange the circles in order, say, **Trace the outside of each of the circles with your finger.**

- **Tell/show me what circle is the biggest around.**
- **Tell/show me what circle is the smallest around.**

Next, using a different color of yarn for each circle, have students run the yarn around the circumference and cut the yarn. Then line up the pieces of yarn to see if students' predictions were correct. If appropriate, have students measure the pieces of yarn with a ruler.

Lesson Follow-Up

If students have difficulty completing this activity, have them complete Activity 34.

Activity 32

Independent **Objective**

Students measure perimeter of polygons.

Have students draw rectangles on one-inch grid paper.

- **How many squares are in your rectangle?**
- **Who has the most squares? The fewest?**

Discuss how the rectangles with more squares take up more space than those with fewer squares.

- **If you wanted to put a fence around your rectangle, how long would it have to be?**

Guide students to see that they would have to count the number of the squares' sides around the edge of the rectangle. Now show students several pre-cut shapes. Tell them to put a "fence" around the shape.

- **How could you find out how long it would have to be?**

Hand each student a ruler. Demonstrate how to use the ruler if necessary. Say, **Here is a ruler. Measure how long the fence will be.**

After students have measured the perimeter with a ruler, have them cut a piece of yarn the same length as the "fence." Have students place the piece of yarn around the edge of the rectangle to see if they were correct.

Repeat the activity with various polygons.

Lesson Follow-Up

If students have difficulty completing this activity, have them complete Activity 35.

Conceptual Development Activities

Activity 33

Independent **Objective**

Students measure capacity using cups, pints, quarts, and gallons.

Gather containers in cup, pint, quart, and gallon sizes. Tape the size name on each container. Say: **Here are containers for liquids.** Give examples of liquids if needed**. Each is a different size.**

Give the name for each one. Let the students explore the containers. Use the name whenever possible. Have students line up the containers from smallest to largest. Ask questions such as:

- **Why do you think the gallon is larger than the quart?**
- **Where should the pint go?**

Tell students that they will be filling up the gallon container using the other containers.

- **How many quarts will it take to fill the gallon container?**
- **Will it take more cups or quarts?**

Use the quart to fill the gallon container with water. Count how many quarts it takes to fill it. Continue this procedure with the pint and cup. Compare the results.

Lesson Follow-Up

If students have difficulty completing this activity, have them complete Activity 36.

Activity 34

Developing **Objective**

Students identify the circumference of circles and compare the area of circles.

Prepare several large tag board circles, and place them on the floor. Ask students to identify what shape they are.

- **Show/Tell me which circle is the largest.**
- **Show/Tell me which circle is the smallest.**

Direct students to place the circles on top of each other so that the largest circle is on the bottom and the smallest is on the top. Now demonstrate how to walk around the outside of the circle toe to heel. Tell students they will be counting how many footsteps it takes to go around the circle.

- **Which circle will take the most footsteps?**
- **Which circle will take the fewest footsteps?**

Each student will then walk around each of the circles, counting the number of footsteps. Students unable to walk can be the step counters. Record the results, and compare them to students' predictions. Let students decorate the inside (area) of the circle emphasizing the difference between the inside (area) and the outside (circumference).

Lesson Follow-Up

If students have difficulty completing this activity, have them complete Activity 37.

Conceptual Development Activities

Activity 35

Developing **Objective**

Students measure the lengths of sides of rectangles and triangles and compare the areas of rectangles and squares.

Give students color tiles, and have them create rectangles.

- **How many tiles are in your rectangle?**
- **Who has the biggest rectangle? The smallest?**
- **Who has the most tiles? The fewest?**

Compare the sizes of the rectangles by counting the number of tiles. Next, have students measure the sides of their rectangles with a ruler or by building a "fence" around their rectangles, putting an extra layer of tiles along the edge.

- **How many of the tiles' sides did it take to build your fence?**

Show students several shapes of different sizes made out of tag board. Make sure the sides are measured to the whole inch. Demonstrate how to measure the sides using the color tiles. Let the students measure the sides to determine which triangle has the longest perimeter. The students can also compare the areas of the different sizes of triangles by placing one on top of another.

Lesson Follow-Up

If students have difficulty completing this activity, have them complete Activity 38.

Activity 36

Developing **Objective**

Students measure capacity using cups.

Introduce the measuring cup. Ask students what it is used for. Show them some other containers for liquids. Ask them to put the containers in order from smallest to largest.

- **Which container will hold the fewest cups?**
- **How many cups do you think it will take to fill it?**
- **Why do you think that?**

Guide them to fill the smallest container using the measuring cup. Record the results. Choose another container to fill.

- **Will this take more cups or fewer cups to fill? Why?**
- **How many cups do you think it will take to fill this container?**

Follow this procedure until all the containers have been filled. Have students compare the size of the containers with the number of cups it took to fill them.

Lesson Follow-Up

If students have difficulty completing this activity, have them complete Activity 38.

Conceptual Development Activities

Activity 37

Emerging Objective

Students recognize the outside (circumference) and the inside (area) of a circle.

Have students stand in a circle. Say, **Here is a circle. Walk around the outside of the circle.**

Have each student walk or move around the circle.

Direct the students to step inward one step to make the circle smaller. Ask:

- **Tell me, is the circle smaller or bigger?**

Have them walk around the outside of the circle. Then direct them to take two steps back, and repeat the question. Again, have them walk around the outside of the circle. Step inside the circle. Tell them they are outside the circle and you are inside the circle. Have each student take a turn stepping inside the circle. Ask each student each time:

- **Are you outside the circle or inside the circle?**

Next, give them a paper with a circle on it. Direct them to trace the outside of the circle with a marker and color the inside with a crayon.

Lesson Follow-Up

If students have difficulty completing this activity, find circular objects such as a plate. Have them run their finger around the outside of the plate and place their whole hand on the inside of the plate. Reinforce the concept of *inside* and *outside* a circle whenever appropriate.

Activity 38

Emerging Objective

Students recognize the outside (perimeter) and the inside (area) of rectangles and triangles.

Make several triangles and rectangles out of tag board. Say, **Here is a triangle. Trace the outside with your finger.** Model while you are saying this.

Give the students a piece of paper and a crayon. Say, **Now trace the outside of the triangle on the piece of paper.**

Repeat with other triangles and rectangles.

Next, place a tag board triangle in front of each student. Also, give each student an object such as a colored counter. Say:

- **Place the colored counter inside the triangle.** (Again model as you are saying this.)
- **Place the colored counter on the outside of the triangle.**

Repeat several times until students demonstrate they understand the difference between the inside and outside of the triangle. Repeat the process with the rectangles.

Lesson Follow-Up

If students have difficulty completing this activity, find familiar objects that are rectangles or triangles such as blocks, chalkboard erasers, or a book. Have them trace around the edge of the object with their finger and then place their hand on the inside of the object.

Conceptual Development Activities

Activity 39

Independent **Objective**

Students express, represent, and use whole numbers to 200.

Supply a container of counters or similar items of various colors. Have approximately 100 counters in the container. Have students guess or estimate how many counters there are. Record their responses. Using bowls, have students separate the counters by color. In pairs or small groups, have students count how many counters of that color there are. Have each group count and record each of the different colors. Then have the groups compare their results. Have them re-count together to solve any discrepancies. Using calculators, have the students add all the color totals to determine how many counters there are altogether. Compare students' results. Find out which student had the closest guess or estimate. Repeat the activity using different items and different quantities to 200.

Lesson Follow-Up

If students have difficulty completing this activity, have them complete Activity 44.

Activity 40

Independent **Objective**

Students identify the value of money to $2.00 expressed as a decimal.

Review with students how to express the values of a penny, nickel, dime, quarter, and dollar bill as decimals. Display a dollar bill and a dime.

- **Show/Tell me how much money this is altogether.**
- **Show me how to write it using a decimal.**

Demonstrate as needed. Continue, increasing the level of difficulty in amounts up to $2.00. Try the reverse. Give each student the necessary coins and bills. Write a dollar amount on the board, such as $1.55. Say:

- **Show/Tell me what bills and coins you need to make $1.55.**

Lesson Follow-Up

If students have difficulty completing this activity, have them complete Activity 45.

Conceptual Development Activities

Activity 41

Independent Objective

Students compare fractional parts including halves, fourths, thirds, and eighths.

Give each student 5 dessert-sized paper plates. Demonstrate how to divide the plates into the following fractions: 1, $\frac{1}{2}$, $\frac{1}{3}$, $\frac{1}{4}$, and $\frac{1}{8}$ using a pencil and a ruler. Check to make sure all the students have completed the task correctly. (If necessary, divide the plates into sections beforehand.) Instruct students to write the appropriate fraction on each section. Have students cut the plates into the fractional pieces they traced and labeled. Have the students compare different fraction sizes. Ask questions such as:

- **Which fraction will you write on this piece? How do you know?**
- **Show/Tell me which fraction is bigger/larger.**
- **Show/Tell me which fraction is smaller.**
- **Show/Tell me how many $\frac{1}{8}$ pieces equal $\frac{1}{2}$.**

If necessary, have students stack the fractional pieces on top of each other to determine how many smaller pieces are needed to make a larger piece.

Lesson Follow-Up

If students have difficulty completing this activity, have them complete Activity 46.

Activity 42

Independent Objective

Students solve two-step real-world problems involving addition and subtraction of two-digit numbers and check for accuracy using the reverse operation.

Give each student a calculator. Set up a scenario for the students such as a group of friends at the beach. Write the addition problem on the board as you say, **Bob found 12 shells at the beach. Sue found 21 shells. Together they found 33 shells.** Show students how to solve on the calculator.

Explain that to check the answer, students will do the reverse operation. Write $33 - 12 = ?$ on the board. Show students how to solve the problem using the calculator. Write the answer. Explain to the students that if you take away (subtract) the number of shells that Bob found, you are left with the number of shells that Sue found. Create different scenarios for the students to solve. After the students show mastery of the skill, introduce another element to create a two-step problem such as this: Bob found 12 shells and Sue found 21 shells at the beach, but on their way home, 6 shells broke. Demonstrate how to solve and check the two-step problem.

- **What numbers do you know?**
- **What number do you want to find out?**
- **Will you add or subtract to find the number?**

Lesson Follow-Up

If students have difficulty completing this activity, have them complete Activity 47.

Conceptual Development Activities

Activity 43

Independent Objective

Students round to the nearest ten to estimate whole numbers to 100.

Review with students how to round to the nearest ten. Point to the number 18 on a number line, and say, **Here is the number 18. It is between 10 and 20 on the number line.**

- **Is it closer to 10 or closer to 20?**

Point to the number 32 on a number line, and say, **Here is the number 32. It is between 30 and 40 on the number line.**

- **Is it closer to 30 or closer to 40?**

Next, give students problems to solve. Have them first estimate an answer, then compute the exact answer. For example, write $19 + 12 = ?$ on the board. Say, **First we are going to estimate our answer**

- **Nineteen is between what numbers on the number line? Is it closer to 10 or closer to 20? Twelve is between what numbers on the number line? Is it closer to 10 or closer to 20?**

Write $20 + 10 = ?$ on the board and have students provide the answer. Have students use calculators to compute the answer to the original problem.

- **Is our estimate close to the answer?**

Work several more examples with students. When they are comfortable with the process, give them addition and subtraction problems to work on their own.

Lesson Follow-Up

If students have difficulty completing this activity, have them complete Activity 48.

Activity 44

Developing Objective

Students express, represent, and use whole numbers to 50.

Supply different quantities of three different objects, such as 23 cars, 15 crayons, and 30 blocks. Also supply number cards that show the numeral, number name, and picture sets representing the number. Have students separate the items. Have students help you count how many are in each set. Ask:

- **Show/Tell me how many cars there are.**
- **Show/Tell me how many crayons there are.**
- **Show/Tell me how many blocks there are.**

When students have identified the quantities, help them match the corresponding number card to each quantity. Demonstrate how to write the amounts as numerals and number names.

Throughout the school day, have the students count as often as possible and write the corresponding numbers.

Lesson Follow-Up

If students have difficulty completing this activity, have them complete Activity 49.

Conceptual Development Activities

Activity 45

Developing **Objective**

Students identify the value of coins to $0.50 expressed as a decimal.

Give each student a penny, nickel, dime, and quarter. Review the names and values of each coin.

- **Show me your penny.**
- **How much is a penny worth?**

Demonstrate writing 1¢. Say:

- **There is another way to write 1 cent.**

Demonstrate writing $0.01. Continue the process using the nickel, dime, and quarter. To show mastery, have students write the decimal for each coin as you hold it up. Then have students write the value of different coin combinations up to $0.50.

Lesson Follow-Up

If students have difficulty completing this activity, have them complete Activity 49.

Activity 46

Developing **Objective**

Students compare and order whole number to 50.

Cut out the shape of a foot about 10 inches long. Make 50 "feet." Have students help you with this if appropriate. Write a numeral and the matching number word on each foot to 50. Starting with the feet numbered 1 through 10, mix up the feet and have the students put the feet in numerical order on the floor. Time them for the extra challenge of beating their previous time, or have groups compete against each other. As the students become more proficient, increase the number of feet used.

Have students practice ordering whole numbers in other ways as well. For example, show students five sets of classroom objects with quantities from 1 to 50 in each set. (You could use books, counters, a jar of pennies, and so on.) Have students count the number of objects in each set, say the number name, write the number, and order the sets. Have students follow the same procedure with pictures of objects or with sets of number cards. As students work, ask:

- **How many are in this set?**
- **How do you write that number?**
- **Is that number bigger or smaller than the first set? How do you know?**

Lesson Follow-Up

If students have difficulty completing this activity, have them complete Activity 50.

Conceptual Development Activities

Activity 47

Developing **Objective**

Students solve addition and subtraction problems with sums to 50 using tallies.

Give each student a dry erase board and marker. As you demonstrate, say, **Let's make 12 tally marks. Now let's make 4 tally marks.**

- **Show/Tell me how many tally marks did we make altogether?**

Count the tally marks together as a class. Repeat using different quantities to 50. As the students become more proficient, show them how to subtract by crossing off tally marks.

Next, give students real-world addition and subtraction problems to solve that involve quantities up to 50.

- **To get ready for Game Day at school, I got 6 soccer balls, 12 tennis rackets, and 8 tennis balls. How many items did I get altogether?**
- **I took 30 sandwiches to a picnic. People ate 17 of the sandwiches. How many sandwiches were left over?**

Have students use groups of counters or tallies to help them solve the problems.

Lesson Follow-Up

If students have difficulty completing this activity, have them complete Activity 51.

Activity 48

Developing **Objective**

Students apply the concepts of counting and grouping to identify the value of whole numbers to 50.

Place 15 crayons on a desk or table. Count 5 crayons and place them in a set; then continue grouping the remaining crayons into sets of 5.

- **How many sets of crayons are on the table?**
- **How many crayons are in each set?**
- **How many crayons are there altogether?**

Remind students how to use skip counting or repeated addition to find the total number of crayons. Repeat the activity with quantities up to 50, having students group different quantities into sets of different sizes. For example, have students group a total of 12 crayons into sets of 2 or group 50 crayons into sets of 10. If students have an incomplete group at the end, tell them to count on to add that to their total.

Next, place 50 crayons on the table. Pick a number, and have students count and group crayons to match that number. Example:

- **I'm thinking of the number 25. Show me 25 crayons.**

Repeat with different quantities up to 50.

Lesson Follow-Up

If students have difficulty completing this activity, have them complete Activity 51.

Conceptual Development Activities

Activity 49

Emerging — Objective

Students match two or more objects to identical objects using one-to-one correspondence.

Draw a line dividing a piece of construction paper in half. Display a classroom object, such as an eraser, on one side of the paper. Place several additional erasers on the desk. Say, **Here is one eraser.**

- **Show me the eraser.**

Say, **I want to match this eraser with another one.**

- **Show me one more eraser.**

Place the additional eraser on the empty half of the paper, aligned with the first eraser. Repeat the activity, starting with two erasers on one side of the paper. Help students match the erasers one-to-one. Continue the process with quantities up to 6.

Lesson Follow-Up

If students have difficulty completing this activity, use objects meaningful to the students. Have two objects displayed in containers that the students can see. Hand the students a like object. Identify the object. Ask the students to put it with the other one. Use the concept of matching and one-to-one correspondence whenever possible throughout the day.

Activity 50

Emerging — Objective

Students compare the size of parts of objects to the whole to determine which is the largest or smallest.

Display a fake or real plant with flowers. Talk about the whole plant and about the parts of the plant, such as petals, leaves, and stem(s). Take off one petal. Say, **This petal is *part* of the plant.**

- **Tell me which is smaller, the petal or the whole plant?**

Take a leaf off the plant.

Say, **This leaf is *part* of the plant.**

- **Tell me which is larger, the leaf or the whole plant?**

Repeat this process with different parts of the plant.

Lesson Follow-Up

If students have difficulty completing this activity, use an item that is meaningful to each student. For example, use a large toy car, and discuss how a wheel is part of the car. Ask students to tell you if the wheel is larger or smaller than the car. Reinforce the concepts of the whole being larger than the part whenever possible.

Conceptual Development Activities

Activity 51

Emerging **Objective**

Students solve simple problems involving joining or separating objects to 6.

Give each student four items that link together, such as linking chains or beads. Each student should have two in one color and two in a different color (for example, two red and two green beads). Show students how to combine the objects of the same color.

- **Show/Tell me how many red beads you have.**
- **Show/Tell me how many green beads you have.**

Have the students join the two sets.

- **Show/Tell me how many beads you have altogether.**

Repeat this procedure, working up to sets of 6. Reverse the process. Start with all objects together.

- **Show/Tell me how many beads you have altogether.**

Have the students separate the two colors.

- **Show/Tell me how many red beads you have.**
- **Show/Tell me how many green beads you have.**

Lesson Follow-Up

If students have difficulty completing this activity, use objects that appeal to the individual students and are meaningful to them. Use the concepts *joining together* and *taking apart* whenever possible; for example, give the students two square crackers and three round crackers. Have students count each set with you and determine how many crackers they have altogether.

Activity 52

Independent **Objective**

Students identify the categories with the largest and smallest numbers represented on a bar graph.

Tell students they will be making a bar graph to show what kind of pets they have. (You may choose to graph other information that is relevant to your students.) First, demonstrate how to compose a graph. Put a title on the graph. Write the different pet types along the *x*-axis and numbers along the *y*-axis. Using a different-color sticky note for each kind of pet, place one sticky note on the graph for each pet. When all the pet types are on the graph, ask:

- **How many cats?**
- **How many dogs?**

Now have students transfer the information to their own personal graphs on one-inch grid paper. They can color the bars with the same color as the sticky notes.

- **Which pet do we have the most of?**
- **Which pet do we have the fewest of?**
- **How do you know?**
- **Are there more cats or dogs?**

Lesson Follow-Up

If students have difficulty completing this activity, have them complete Activity 53.

Conceptual Development Activities

Activity 53

Developing **Objective**
Students identify the category with the largest number in a pictograph.

Give each student a handful of three kinds of pattern blocks. For example, they may have four diamonds, two squares, and five triangles. Give them grids sized so the pattern blocks will fit in the squares. Have students sort the pieces and place them in lines on the grid paper. Make sure they put one piece in each square.

- **Which line is the longest?**
- **What did you have five of?**
- **Which shape did you have the most of?**
- **Are there more squares or triangles?**

Repeat this activity several times with different amounts of blocks. If appropriate, have students trace the blocks on the grid paper and color.

Lesson Follow-Up
If students have difficulty completing this activity, have them complete Activity 54 (for MA.6.S.6.1) or Activity 55 (for MA.6.S.6.2).

Activity 54

Emerging **Objective**
Students identify the largest set of objects, pictures, or symbols to 6 representing data in an object graph or pictograph.

Demonstrate how to sort color tiles by color and put them on one-inch grid paper. Give each student six color tiles. (Use only two colors, and make sure students have different quantities of each color.) Have a number line nearby. Help students sort the tiles and place them on the paper.

- **How many red tiles do you have?**

Say, **Find that number on the number line.**

- **How many blue tiles do you have?**

Say, **Find that number on the number line.**

- **Do you have more red tiles or more blue tiles?**

If appropriate, have students color in the grid paper to match the color tiles. Repeat this activity several times with different amounts of each color tile.

Lesson Follow-Up
If students have difficulty completing this activity, create graphs with two tiles (one red and two blue). Use sets of objects to reinforce the concept of *more* throughout the day.

Activity 55

Emerging **Objective**

Students identify the largest set of objects, pictures, or symbols in an object graph or pictograph.

Use two colors of teddy bear counters. Show students how to sort them by color. Give six counters to each student, and have students sort them by color. (Make sure they do not have the same amount of each color.) Show them how to line up the counters.

- **Which line is the longest?**
- **Which color has the most?**
- **How many does that color have?**

Find the number on the number line. Now count the other color. Show students that there is more of one color than the other. Repeat this activity several times using different amounts of counters.

Lesson Follow-Up

If students have difficulty completing this activity, use objects they are familiar with and that are similar in size, such as toy cars or blocks. Line up the objects side-by-side, and have students show or tell you which group has more.

NOTES

MIDDLE SCHOOL

Conceptual Development Activities

Activity 1

Independent Objective

Students solve problems involving ratios using models and charts.

Show the students a picture of a s'more and explain the construction of the item: 2 graham crackers, 1 marshmallow, and 1 piece of chocolate. Demonstrate, on a white board or overhead projector, how to represent the ingredients needed for one person to make the snack. The representation could be a stick figure with 2 squares below it for the graham crackers, 1 circle for the marshmallow, and 1 rectangle for the chocolate. Ask the students to copy these figures. Then ask the students to create the figures again in order to make 2 s'mores. Ask the students how many graham crackers are needed for 2 people. How many marshmallows for 2? This could be repeated for 3 or more people.

- **If you need 2 graham crackers for 1 person, show/tell me how many you need for 2 people.**
- **If you need 2 graham crackers for 1 person, show/tell me how many you need for 3 people.**
- **If you need 1 marshmallow for 1 person, show/tell me how many you need for 3 people.**

Lesson Follow-Up

If students have difficulty completing this activity, have them complete Activity 4.

Activity 2

Independent Objective

Students identify that a higher percent represents a larger quantity or amount.

Use a catalog for the students to choose an item they would like to "purchase." Distribute created "coupons" signifying a percentage amount the items can be discounted. The coupons can be 10% off or 50% off of regular price. The coupons should have various percentages, sizes, shapes, and colors to ensure students are focusing on the number and not on the size or color of the coupon. Ask the students which coupon would save the most money when buying the item of their choice. This will help students recognize that a higher percentage discount will result in more money saved.

- **Show/Tell me where the percent discount is on the coupon.**
- **Show/Tell me which coupon will give you the largest percent discount.**
- **Show/Tell me which coupon will save you the most money.**

Lesson Follow-Up

If students have difficulty completing this activity, have them complete Activity 5.

Conceptual Development Activities

Activity 3

Independent **Objective**

Students measure and describe how models compare in size to real-life objects.

Have several toy cars, enough for each group of two students. Have students measure the diameter of the wheels, the width of the hood, and the length of the car in centimeters. Record their results on a sheet of paper. Then take a tape measure and go outside and measure the same measurements on a real car. Record these measurements in meters. Help the students convert their measurement of the real car from meters to centimeters.

- **Show/Tell me which is larger.**
  a. toy car b. real car
- **Show/Tell me which could be the diameter of the tire of a real car.**
  a. 2 cm b. 20 cm c. 200 cm
- **Show/Tell me which could be the width of the hood of a toy car.**
  a. 4 cm b. 400 cm c. 4,000cm

Lesson Follow-Up

If students have difficulty completing this activity, have them complete Activity 6.

Activity 4

Developing **Objective**

Students solve problems involving simple ratios using objects or pictures.

Demonstrate the construction of a s'more by using pictures of the ingredients or the ingredients themselves. Assist the students in constructing their own. Ask the students to show how many graham crackers they need to make one s'more by holding up the pictures or the ingredients. Then, ask the students to construct a second s'more. Ask the students to show how many crackers are needed to make two s'mores.

Have the pictures or ingredients available for the students to use as they solve the problems.

- **If you need 2 graham crackers for 1 person, show/tell me how many you need for 2 people.**
- **If you need 2 graham crackers for 1 person, show/tell me how many you need for 3 people.**
- **If you need 1 marshmallow for 1 person, show/tell me how many you need for 3 people.**

Lesson Follow-Up

If students have difficulty completing this activity, have them complete Activity 7.

Conceptual Development Activities

Activity 5

Developing Objective

Students identify that percent discounts reduce the price of goods.

Bring in coupons from newspapers or the internet. Show students the percentage discount on the coupons. Explain that students will pay less money for an item if there is a larger percentage discount. Also bring in a sales flyer and ask the students to circle with their pencil or finger the discounts they find. This demonstrates that the student has the ability to distinguish the percentage discount from other print features on the flyer. Create a coupon for percentage discount on an item familiar to the student. Also create a similar piece of paper with no discount on it.

- **Show/Tell me which paper you would use to buy the item for less money.**
- **Show/Tell me the percentage on the coupon.**
- **Show/Tell me the percentage on the flyer.**

Lesson Follow-Up

If students have difficulty completing this activity, have them complete Activity 8.

Activity 6

Developing Objective

Students compare the size of models to real-life objects using *same, larger,* and *smaller.*

Find some toy cars, a large plastic ant or other insect, toy elephant, and other models of real-life objects. Hold up the toy car and ask if it is larger or smaller than a real car. After students answer "smaller," explain that models look like another object but they don't usually work like the real object and are usually not the same size as the real object. The car model was smaller, but some models can be larger than the real object. Hold up the plastic ant. Ask if it is a model of an ant. When the students say "yes," ask them if it is larger or smaller than a real ant.

- **Show/Tell me if this model is larger or smaller than a real car.** (Hold up toy car.)
- **Show/Tell me if this model is larger or smaller than a real ant.** (Hold up plastic ant.)
- **Show/Tell me if this model is larger or smaller than a real elephant.** (Hold up toy elephant.)

Lesson Follow-Up

If students have difficulty completing this activity, have them complete Activity 8.

Conceptual Development Activities

Activity 7

Emerging Objective

Students solve simple problems involving a 2 to 1 ratio using objects.

Explain that each student will receive two cookies or two pictures of cookies. Give two cookies or pictures to the first student. Repeat with the second student. Ask the students to count 1-2 for the number of students. Ask the students to count how many cookies were needed for these two students. Repeat for the number of students available.

- **Show/Tell me how many cookies we need for one person.**
- **Show/Tell me how many cookies we need for 2 people.**
- **Show/Tell me how many cookies we need for 3 people.**

Lesson Follow-Up

If students have trouble acquiring the concepts, go back and work on counting to 6 and one to one correspondences. Use prompting to help the student count. Then return to the activity. Use different objects to help generalize the concept.

Activity 8

Emerging Objective

Students match objects to a model or picture that is a smaller version.

Identify five items with which the students are familiar. These items must be small enough to fit on a table such as a shoe, a book, and so on. Take pictures of each of the five items. Show the pictures to the students. Have or help the students point to the pictures of the object and say the name of the object. Have or help the students place the picture on the object. Repeat for each object individually. Place three of the items on the table and ask the students to match the picture to the correct item.

- **Show/Tell me which object you see here.**

Lesson Follow-Up

If students have difficulty with this activity, reduce the number of objects from which the students have to choose, or use hand-over-hand prompts to assist.

Conceptual Development Activities

Activity 9

Independent Objective

Students identify properties of three-dimensional figures including pyramids, prisms, or cylinders.

Have examples of pyramids and prisms. Put the prisms in one pile and the pyramids in another pile. Ask students to tell you some things each of the piles have in common. Possible answers may include: they are three-dimensional or each one has a base. Characteristics that the pyramids share may include: they each have one base, or the sides end in a point. Characteristics that the prisms share may include: they each have two bases, they have parallel sides, or the two bases are identical. If students do not come up with these characteristics, help them discover the answer by doing some of the following: holding the shapes and showing the bases, touching the points of the pyramids, or asking them what the top of the pyramids look like. Pull out a cone and a cylinder. Help the students decide in which pile the cone should be based on how the top looks. Do the same with the cone.

- **Show/Tell me a property of a pyramid.**
- **Show/Tell me a property of a prism.**
- **Show me a prism.** (Choose from one of three shapes.)

Lesson Follow-Up

If students have difficulty completing this activity, have them complete Activity 11.

Activity 10

Independent Objective

Students use stated formulas to solve for perimeter and area of rectangles.

Draw a picture of a rectangle on the board. Label the width as 3 inches and the length as 5 inches. Give a student a colored marker (e.g., green) and have the student go up and draw over two sides of the rectangle that are the same length. Have another student come up and draw over the remaining two sides that are also the same length in a different color (e.g., red). Ask the student how long each of the green sides are. When they answer "3" write 3 on the board. Ask the students how many sides are 3 inches long. When they answer "2" add $\times$ 2 behind the 3. Ask the students to solve the equation to find the length of both green sides. Repeat for the red sides. Then ask the students to add the total of the two equations to find the perimeter of the whole shape. Repeat this several times with different size rectangles.

Draw a rectangle that is 4 inches by 8 inches.

- **Show/Tell me which two sides are equal.**
- **Show/Tell me which equation is the equation that shows the length of the two long sides of the rectangle.**
  a. $3 \times 8 = 24$ **b.** $4 \times 8 = 32$ **c.** $2 \times 8 = 16$
- **Show/Tell me which equation is the equation that shows the perimeter of the given rectangle.**
  a. $(2 \times 8) + (2 \times 4) = 24$
  b. $8 \times 4 = 32$
  c. $8 \times 8 \times 4 \times 4 = 1{,}024$

Lesson Follow-Up

If students have difficulty completing this activity, have them complete Activity 12.

Conceptual Development Activities

Activity 11

Developing Objective

Students identify three-dimensional figures, including pyramids, prisms, cylinders, and cones.

Hold up a prism and show students both bases. Hold up a pyramid and show students the base and the point. Now, hold up a shape and ask if it has two bases or one base and a point. Tell students that if it has a point it is a pyramid. If it has two bases it is a prism. Hold up various shapes and have students tell you if it is a prism or a pyramid. If students aren't sure, cue them by asking how many bases the shape has, or if the shape has a point. Repeat this several times with different pyramids and prisms. To introduce a cylinder and a cone tell students that shapes with a circle for a base are special and have special names. A cylinder is a special type of prism. A cone is a special type of pyramid.

Show a pyramid.

- **What is this object?**
  a. cone   b. pyramid   c. prism

Show a prism.

- **What is this object?**
  a. cone   b. pyramid   c. prism

Lesson Follow-Up
If students have difficulty completing this activity, have them complete Activity 13.

Activity 12

Developing Objective

Students add lengths of sides of rectangles to measure the perimeter and find the area in square units.

Draw a picture of a rectangle on the ground either by using sidewalk chalk outside or tape on the floor in the classroom. Draw the square units inside the rectangle as if it were drawn on graph paper. Each square should be about 1 square foot. Have students start at a corner of the rectangle. Each student should have a piece of paper with a rectangle drawn on it and a clipboard. Have the students walk one side of the rectangle counting each unit as they walk. When the students reach a corner they should write the length of that side of the rectangle on their papers. When the students have walked the entire rectangle they should add up the four numbers they wrote on their papers to find out how far they walked around the rectangle. Repeat this with several different-sized rectangles. Then repeat this exercise with rectangles drawn on graph paper and students using their fingers to "walk" around the rectangle.

Draw a rectangle that is 4 inches by 8 inches on graph paper.

- **Show/tell how long each side is.**
- **Show/Tell me how far you would walk around two sides of the rectangle.**
- **Show/Tell me what the perimeter of the given rectangle is. How far around the rectangle do you have to walk?**

Lesson Follow-Up
If students have difficulty completing this activity, have them complete Activity 14.

Conceptual Development Activities

Activity 13

Emerging **Objective**

Students recognize common three-dimensional figures such as spheres, cubes, cylinders, or cones.

Hold up different-sized balls and tell the students that balls are also called spheres. Find other spheres such as marbles and gumballs and tell the students that these objects are also spheres. Have students describe spheres. Put a ball in the middle of two other objects that are not spherical. Ask the students to show you the sphere. Repeat this with other spherical objects. Then hold up a number cube and tell students that this is called a cube. Find other cubes such as a box, or a puzzle cube, and tell the students they are also cubes. Have students discuss the properties of cubes. Put a number cube in the middle of two other objects that are not cubes. Ask the students to show you the cube. Repeat this with other cubes. Similar activities can be done with cylinders and cones.

Lesson Follow-Up

If students have difficulty completing the activity, hold a ball and say "this is a sphere in math class." Have the students point to the sphere using hand-over-hand assistance if necessary. Repeat this until the students can point to the sphere independently. Do not switch to different spheres until the students can label the original sphere independently.

Activity 14

Emerging **Objective**

Students match common three-dimensional figures that are the same size.

Have 2 different types of soft drink cans on the table in front of the students. Hold up one of the cans and ask the students to point to something that is the same size and shape as the can being held up. When the students can reliably point to the can on the table, add another object to the table that is a different shape and size such as a granola bar. Hold up one of the soft drink cans and ask the students to point to the object that is the same shape and size. When the students can reliably choose the other soft drink can, add an object that is the same shape but smaller such as one of the half-size soft drink cans. Repeat the above activity. As the students become proficient add more non-example objects.

Have a soft drink can and one other object on the table.

- **Show/tell me which object on the table is the same size and shape as the item I am holding.**

Have a soft drink can and two other objects on the table.

- **Show/tell me which object on the table is the same size and shape as the item I am holding.**

Lesson Follow-Up

If students have difficulty completing the activity, return to having just one object on the table that is the same size and shape as the object being held. Use hand-over-hand to help the students point to the object if necessary. Use the same brand of soft drink to begin with so the cans are identical. When the students can successfully and independently complete the activity with the same soft drink can, move to using different brands of soft drinks.

Conceptual Development Activities

Activity 15

Independent **Objective**

Students solve equations involving addition and subtraction of numbers to 500.

In a class of 6 students give each student a number cube and a calculator. If you have more students, put them in groups so that you have six groups. Write the following on the board: __ __ + __ __ + __ __ =______
Have the first student role the dice and write the number in the first space. Have the second student role the dice and write the number in the second space. Continue for all the spaces. When the equation is written on the board, tell the students to solve the equation using their calculators, but that they can't start solving until you say "go." Count "1, 2, 3, . . . go." The first student/group to get the correct answer gets a point. Play several rounds of the game. If one student/group is consistently winning, let them be the teacher and write the numbers on the board and the teacher can role the dice for that group. Variations to the game may include writing the following equation for subtraction with groups of 5: __ __ __ − __ __ =_______ .

Ask students the following:

- **Show/Tell me the correct answer for the following equation: 356 + 72 = ______**
- **Show/Tell me the correct answer for the following equation: 23 + 299 = ______**
- **Show/Tell me the correct answer for the following equation: 257 − 98 = ______**

Lesson Follow-Up

If students have difficulty completing this activity, have them complete Activity 19.

Activity 16

Independent **Objective**

Students solve equations involving basic multiplication and division facts.

Make a set of cards with multiplication facts on them. Write a multiplication fact on the first card. On the next card put the answer to the first equation and beneath it write a new multiplication fact. Continue this pattern until you have one more card than you do students. On the first card, write the answer to the last equation. Hold onto the first card. Shuffle the rest of the cards and pass them out to the students. Read the first equation. The student with the answer says the answer and reads the equation on his or her card. This continues until the last equation is read and you have the answer. Once students understand how this game is played, time can be kept for how fast the class can complete all the cards. The students can work to complete the sequence in the fastest time possible, or you can set a benchmark time to beat. The same game can be played with division facts. Ensure that no facts with the same answer are used.

- **Show/Tell me the correct answer for the following number sentence: 5 × 7 = ______**
- **Show/Tell me the correct answer for the following equation: 6 × 4 = ______**
- **Show/Tell me the correct answer for the following equation: 3 × 4 = ______**

Lesson Follow-Up

If students have difficulty completing this activity, have them complete Activity 20.

Conceptual Development Activities

Activity 17

Independent **Objective**

Students translate real-world problem situations into equations involving addition and subtraction of two-digit numbers.

Explain to students that you are code breakers working for the government. You have to translate English into a number so that the enemy cannot decipher your plans. Write the first word problem on the board: *Lisa's class checked out fifteen books from the library in one week. Jim's class checked out nineteen books in the same week. How many books were checked out altogether?* Read the first sentence aloud.

- **I know how to translate *fifteen* into math. Who else knows?**

Write the number 15 under the word *fifteen*. Do the same for the word *nineteen*. Read the last sentence aloud. Say: **I know that *how many* means I'm looking for a number.** Put a question mark under those words.

- **I know that *altogether* often means addition. Where should I put the addition sign?**

Place the addition sign between the 15 and 19.

- **What should go between the 19 and the question mark?**

Place an equal sign between the 19 and the question mark. Repeat with other similar addition problems. This lesson works for addition but can be modified for subtraction.

- **Show/Tell me the correct math equation for the following problem: Sue earned twenty-one dollars. George earned fifty-four dollars. How much money did they earn together?**

Lesson Follow-Up

If students have difficulty completing this activity, have them complete Activity 21.

Activity 18

Independent **Objective**

Students use the property of equality to solve problems.

Ask the class to choose a number between 5 and 15. Make index cards with different equations that equal the number the students choose. For example, if the students choose 12 the cards could include $8 + 4$, $10 + 2$, $14 - 2$, 4×3, and 6×2. Post equal signs on the board and have students put their index cards on either side of the equal sign. So there should be equations that look like this: $6 \times 2 = 8 + 4$ around the board. Talk with the students about how even though there are different numbers, both sides of the equal signs are the same number. Explain that in math this means both sides are equal. Do this exercise with different numbers. Have students create some of the expressions on the index cards.

- **Show/Tell me which expression is equal to 4 + 2.**
  a. 3 + 4 b. 4 + 3 c. 5 + 1
- **Show/Tell me which expression is equal to 10 + 7.**
  a. 10 + 6 b. 21 − 4 c. 14 + 5
- **Show/Tell me which expression is equal to 12 − 7.**
  a. 2 + 3 b. 12 − 5 c. 12 + 7

Lesson Follow-Up

If students have difficulty completing this activity, have them complete Activity 22.

Conceptual Development Activities

Activity 19

Independent **Objective**

Students add and subtract one-digit and two-digit equations.

In a class of 4 students, give each student a die and a calculator. If you have more students, group them so that you have four groups. Write the following on the board:

__ __ + __ __ =______

Have the first student role the die and write the number in the first space. Have the second student role the die and write the number in the second space. Continue for all the spaces. When the equation is written on the board, tell the students to solve the equation using a calculator, but that they cannot start solving until you say "go." Count "1, 2, 3, go." The first student/group to get the correct answer gets a point. Play several rounds of the game. If one student/group is consistently winning, that group becomes the teacher and writes the numbers on the board and you can role the die for that group. Variations to the game may include writing the following equation for subtraction: __ __ __ − __ =________ .

- **Show/Tell me the correct answer for the following equation: 56 + 42 = ______**
- **Show/Tell me the correct answer for the following equation: 23 + 2 = ______**
- **Show/Tell me the correct answer for the following equation: 25 − 9 = ______**

Lesson Follow-Up

If students have difficulty completing this activity, have them complete Activity 23.

Activity 20

Developing **Objective**

Students solve problems that involve combining (multiplying) or separating (dividing) equal sets with quantities to 50 using objects.

Give each student 12 pennies, 6 index cards, and a piece of paper. The paper should have 3 columns. The first column should be labeled *Cards,* the second column labeled *Pennies,* and the third column labeled *Total.* There should be 5 numbered rows. Have the students lay out two index cards. Ask the students to put the same number of pennies on each card and be sure to have no pennies left over. If the students need help getting started, demonstrate how the pennies can be dealt out one at a time to each card. After students have 6 pennies on each of the two cards have them fill out the worksheet. Help them by asking how many cards they have out. Write a 2 under the column labeled *cards.* Ask the students how many pennies are on each card. Write a 6 under the column labeled *pennies.* Ask the students how many pennies they have altogether. Write 12 under *Total.* Repeat with the students using 3, 1, 4, and 6 index cards.

For each question, have the correct number of counters for the students to use.

Have 21 counters.

- **Show/Tell me how many counters would be in each group if you make 3 groups.**
- **Show/Tell me how many counters would be in each group if you make 7 groups.**

Lesson Follow-Up

If students have difficulty completing this activity, have them complete Activity 23.

Conceptual Development Activities

Activity 21

Developing Objective

Students write and solve equations involving addition and subtraction with one-digit and two-digit numbers.

Explain to students that you are code breakers working for the government. You have to translate English into a math code so that the enemy cannot decipher your plans. Write the first word problem on the board: *Lisa's class checked out five books from the library in one week. Jim's class checked out six books in the same week. How many books were checked out altogether?* Read the first sentence aloud.

- **I know how to translate the word *five* into a number. Who else knows?**

Write the number 5 under the word *five*. Do the same for the word *six*. Read the last sentence aloud. Say, **I know that *how many* means I'm looking for a number.** Put a question mark under those words.

- **I know that *altogether* often means addition. Where should I put the addition sign?**

Place the addition sign between the 5 and 6.

- **What should go between the 6 and the question mark?**

Place an equal sign between the 6 and the question mark. Point out that math is often shorter than English. Repeat with other similar addition problems. This lesson works for addition but can be modified for subtraction.

- **Show/Tell me the correct math equation for the following problem: Sue earned one dollar. George earned four dollars. How much money did they earn together?**

Lesson Follow-Up

If students have difficulty completing this activity, have them complete Activity 23.

Activity 22

Developing Objective

Students use physical models to demonstrate the concept of equality.

Give students 2 large index cards and 15 counters. Tell them the two index cards represent a brother and a sister. The brother and sister always have the same amount of counters so they are always equal. Have students put 5 counters on each index card. Say, **The brother was given another counter.** Have the students put another counter on his index card.

- **What does the sister need to keep everything equal?**

Have the students put another counter on the sister's card so that both are equal again. Repeat this with other numbers. Add and subtract counters from the cards.

Have two index cards on the table with 4 counters on each card.

- **Both cards need to be equal. Show/Tell me the correct number of counters you need to add to the other card if I put 2 more counters on this card.**
- **Both cards need to be equal. Show/Tell me the correct number of counters you need to take away from the other card if I take away 3 counters from this card.**
- **Both cards need to be equal. Show/Tell me what you need to do to the other card if I put 3 more counters on this card.**

Lesson Follow-Up

If students have difficulty completing this activity, have them complete Activity 24.

Conceptual Development Activities

Activity 23

Emerging — Objective

Students solve simple problems involving joining or separating sets of objects to 7.

Have a bag of colored counters. Put out a set of 2 colored counters and a set of 1 colored counters. Ask the students to join the two sets and tell you how many counters there are altogether. Help the students push the sets together and count the counters one at time to find the correct answer. Repeat with sets of 1 and 3, 2 and 2, or 3 and 4. Be sure that you explain that the two sets are joined together and you want to find out how many colored counters there are altogether. When the students can independently do this activity, change the activity to having a set of 4 counters and take 2 away. Ask, **How many counters are left?** Help the students take away the 2 counters and count how many are left. Repeat with sets of 3 and 1, 5 and 3, or 7 and 2.

Have a set of 2 counters and a set of 1 counters.

- **Show/Tell me how many counters there are altogether.**

Have a set of 5 counters and ask the student to take 3 counterss away.

- **Show/Tell me how many counters are left.**

Lesson Follow-Up

If students have difficulty completing the activity, start with the joining activity. Have 2 sets with 1 colored candy in each set. Using hand-over-hand, help the students join the two sets. Work on having the students join 2 different sets until the students can do this step independently. Then have the students count the colored candies in the pile. Help the students swith 1-to-1 correspondence.

Activity 24

Emerging — Objective

Students solve simple problems involving small quantities using *more, less, same, larger, smaller,* and *none.*

Have a bag of colored counters. Put out a set of 2 counters, and a set of 7 counters. Circle the set of colored counters that is larger while saying **This is the larger set.** Then ask the students to circle the larger set. Prompt them to get the correct response. Repeat with different sets, but with one set always visually larger than the other. Once the students are doing this independently, change the wording to **Show me the set that has more.**

Have a set of 2 colored counters and a set of 5 colored counters.

- **Show/Tell me which set is larger.**

Have a set of 3 colored counters and a set of 7 colored counters.

- **Show/Tell me which set is larger.**

Have a set of 5 colored counters a set of 3 colored counters.

- **Show/Tell me which set has more.**

Lesson Follow-Up

If students have difficulty completing the activity, go back to the beginning of the activity. Use hand-over-hand prompts if necessary. Use only the word *larger*. Create a visual cue for *larger* if that helps the students. Be sure that one pile is visually larger than the other so that counting is not needed.

Conceptual Development Activities

Activity 25

Independent Objective

Students identify the effects of changes in the lengths of sides of rectangles on the perimeter and area using physical models.

Give students graph paper. Show students a 3-by-3 square. Have them cut out a 3-by-3 square, a 4-by-4 square, and a 5-by-5 square. Have the students create a table with the headings *Length of 1 side, Perimeter,* and *Area.* Help the students fill in the table for the 3-by-3 square. The length is 3, perimeter is 12, and area is 9. If the students can do this easily let them fill in the next two on their own. If they are having trouble, fill the table in as a class.

- **What do you see happening to the perimeter as the length of the side gets bigger?**
- **What do you see happening to the area as the length of the side gets smaller?**

If they need more practice, have them create 2-by-2 squares and 6-by-6 squares.

- **Show/Tell me what happens to the area of a square if the sides become longer.**
- **Show/Tell me what happens to the area of a square as the sides become smaller.**
- **If I have a square with sides that are 4 cm long, is the area larger or smaller than a square with sides that are 2 cm long?**

Lesson Follow-Up

If students have difficulty completing this activity, have them complete Activity 29.

Activity 26

Independent Objective

Students identify examples of slides (translations), turns (rotations), and flips (reflections) of geometric figures.

Create a deck of cards using 8 shapes. Each shape is repeated 4 times, one time in the beginning position, one time rotated 90 degrees, one time flipped over the *x*-axis, and one time translated over the *y*-axis. Have students play "Go Fish" for the cards. The goal is to make a set of 4 shapes. Students must ask for the shape (square) and the way it is moved (translated, rotated, reflected, or beginning). For this, the students may need a worksheet showing the shapes in the beginning position with the shape's name beneath it.

Put 3 cards on the table for the student to touch/point to. They should all be the same shape, but one should be rotated, one reflected, and one translated. You have a picture of the shape in the start position.

- **Show/Tell me which of these three cards on the table is the same shape as the one I am showing but has been rotated.**

Repeat this question for reflected and translated.

Lesson Follow-Up

If students have difficulty completing this activity, have them complete Activity 30.

Conceptual Development Activities

Activity 27

Independent Objective

Students identify common uses of a coordinate place, such as a map or line graph.

Organize students into pairs and give each pair a map. Call out a coordinate location and have students locate that location on the map. This can be played as a game or done as a class exercise. If students have not had much experience with maps, try to find a map of their community and use letters (A, B, C, D) and numbers rather than latitude and longitude for the coordinate lines.

Using the map ask each student to show locations for given coordinates.

- **Show/Tell me what is at D4.**

Continue using different coordinates.

Lesson Follow-Up

If students have difficulty completing this activity, have them complete Activity 31.

Activity 28

Independent Objective

Students convert measures of capacity including cups, pints, quarts, and gallons.

Make 3-by-3 bingo cards. Have the center space be labeled *free*. Write different even quantities of pints and cups on the cards. Make corresponding index cards with quarts and gallons to call off during the bingo game. Put up a chart that shows the gallon, quart, pint, and cup equivalences. Call off a number of quarts, and students with that number of cups or pints put a token on it. The first person with three in a row wins.

- **Show/Tell me how many cups are in a quart.**
- **Show/Tell me how many pints are in a gallon.**
- **Show/Tell me how many pints are in 2 quarts.**

Lesson Follow-Up

If students have difficulty completing this activity, have them complete Activity 32.

Conceptual Development Activities

Activity 29

Developing Objective

Students recognize that changes in the lengths of sides of rectangles will make the figure or object smaller or larger.

Give students graph paper and have them cut out a square with side lengths of 3 units. Ask them if they were to cut out a square with the length of the sides being 4 units, would it be larger or smaller than the square they have? Record their guesses on the board. Then have them cut out a square that is 4 units by 4 units and have them compare it to their first square. Repeat with a 5-by-5 square. Then ask the students if they were to cut out a square with the length of the sides being 2 units by 2 units would it be larger or smaller than their first square? Record their answers. Then have them make the 2-by-2 square. Have them put their squares in a row from smallest to biggest. Ask them to create a rule describing what happens to the size of a square as the sides get longer. Then have them create a rule describing what happens to the size of a square as the sides get shorter.

- **Show/Tell me what happens to a square if the sides become longer.**
- **Show/Tell me what happens to a square as the sides become shorter.**
- **Show/Tell me: if I have a square with sides that are 4 cm long, is it larger or smaller than a square with sides that are 2 cm long?**

Lesson Follow-Up

If students have difficulty completing this activity, have them complete Activity 33.

Activity 30

Developing Objective

Students match congruent figures in different positions, including reflections and rotations.

Create a deck of cards using 8 shapes. Each shape is repeated 4 times, one time in the beginning position, one time rotated 90 degrees, one time flipped over the *x*-axis, and one time flipped over the *y*-axis. Have students play a game like rummy with the cards, the goal being to get a set of 4 cards that are the same shape. This deck of cards can also be used to play a "go fish" type of game. For this, the students may need a worksheet showing the shapes in the beginning position with the shape's name beneath it.

Put 3 cards on the table for the students to touch/point to. One should have a small rectangle, one a small triangle, and one a small square. You should have pictures of each shape rotated 90 degrees or flipped (reflected) over the *x*- or *y*-axis.

Hold up one of the shapes.

- **Show/Tell me which of these three cards on the table is the same shape as the one I am showing you.**

Repeat this question with each shape.

Lesson Follow-Up

If students have difficulty completing this activity, have them complete Activity 34.

Conceptual Development Activities

Activity 31

Developing **Objective**

Students recognize a common use of a coordinate plane, such as a map.

Find a map of the world. Make sure the grid lines are visible. Point the lines out to the students. Team the students up in pairs. One partner is designated as longitude, the other as latitude. Give each student his or her number. The students trace their correct line until they meet and tell you at what country/city they meet. At first use only North and East coordinates. As the students become more proficient you can add in South and West coordinates.

- **Show/Tell me where the 20-degrees north line is on a map.**
- **Show/Tell me where the 20-degrees south line is on a map.**
- **Show/Tell me where the 20-degrees east line is on a map.**

Lesson Follow-Up

If students have difficulty completing this activity, have them complete Activity 35.

Activity 32

Developing **Objective**

Students convert measures of capacity, including the number of cups in a pint and quart.

Make 3-by-3 bingo cards. Have the center space be labeled *free*. Write different even quantities of pints and cups on the cards. Make corresponding index cards with quarts and pints to call off during the bingo game. Put up a chart that shows the quart, pint, and cup equivalences. Call off a number of quarts and students with that number of cups or pints put a token on it. The first person with three in a row wins.

Have the chart available for answering these questions.

- **Show/Tell me how many cups are in a pint.**
- **Show/Tell me how many pints are in a quart.**
- **Show/Tell me how many pints are in 2 quarts.**

Lesson Follow-Up

If students have difficulty completing this activity, have them complete Activity 36.

Conceptual Development Activities

Activity 33

Emerging **Objective**

Students match two- and three-dimensional objects with the same shape but different size.

Create 3-by-3 bingo cards with a free space in the center and 8 shapes around it. Create large pictures of those same 8 shapes on heavy paper. Give each student a bingo card. Hold up a picture of a shape and have students put a token over the same shape on their paper. First student to get three in a row wins.

Put 3 cards on the table for the student to touch/point to. One should have a small rectangle, one a small triangle, and one a small circle. You should have a larger picture of each of the shapes.

- **Show/Tell me which of these three cards on the table is the same shape as the one I am showing you.**

Repeat this question with each shape.

Lesson Follow-Up

If students have difficulty completing the activity, make a bingo card that is 2-by-2. Help the students find the same shape on their card. Make the outline of the shape raised so the students can trace the shape. One way is to do this is to outline the shape with glue and let the glue dry.

Activity 34

Emerging **Objective**

Students recognize objects that have been rotated.

Create a deck of 16 cards with 2 sets of 8 shapes. The second set of cards is each shape rotated 90 degrees. Pair students up and have students play a matching game where they turn over cards and if the cards are the same shape, they keep the pair. If the cards are not the same shape then they turn the two cards over and the next person takes his or her turn.

Put 3 cards on the table for the student to touch/point to. One should have a small rectangle, one a small triangle, and one a small circle. You should also have a picture of each of the shapes rotated 90 degrees.

- **Show/Tell me which of these three cards on the table is the same shape as the one I am showing you.**

Repeat this question with each of the shapes.

Lesson Follow-Up

If students have difficulty completing the activity use a smaller deck of cards and have the students play with the cards turned face up.

Conceptual Development Activities

Activity 35

Emerging **Objective**

Students solve problems using directional or positional language, such as *up, down, left, right,* and *next to.*

Take the students to a grocery store. Have each student be responsible for finding an item. Take the student to the row where that item is found and give the student directions to find the item, such as *it is on the left side, it is next to the cake mix,* or *it is below the ketchup.* If a trip to the store is not possible, then do the same activity with supplies on a shelf in the classroom.

Put 3 objects on the table that are familiar to the students such as a pencil, piece of paper, and an eraser. Lay them on the table in a vertical line.

- **Show/Tell me which item is below the eraser.**
- **Show/Tell me which item is above the paper.**
- **Show/Tell me which item is next to the pencil.**

Lesson Follow-Up

If students have difficulty completing the activity, focus on only one word or phrase, such as *next to.* Do repeated activities where the students need to find the object next to another object. Use hand-over-hand prompting if necessary.

Activity 36

Independent **Objective**

Students identify similarities and differences in features of objects, such as shape and size.

Hold up a red marble and a green marble. Ask students how they are different. Then ask students how they are the same. If the students say they are both marbles, help them to see they are both round. Do the same with a piece of paper and a white cube or die. Ask students how they are different. Then ask students how they are the same.

Hold up two objects.

- **Show/Tell me how they are the same.**
  a. same shape **b. same color** **c. same size**
- **Show/Tell me how they are different.**
  a. different shape
  b. different color
  c. different size

Lesson Follow-Up

If students have difficulty completing the activity, focus on only one characteristic, such as color. Ask, **Is the color the same or different?** Then add another characteristic.

Conceptual Development Activities

Activity 37

Independent **Objective**

Students express, represent, and use percents, including 50% and 100%, and decimals in the context of money to $5.00 or more.

Talk to students about the fact that 50% is the same as half. Half a pizza is 50% of the pizza. Use other concrete examples. Give each student ten 1-dollar play-money bills. Hold up different prizes that students can "buy" for a certain price. When a student chooses to buy something, the student raises his or her hand. If two students want to buy it they then have to out bid each other, as in an auction. The winner then spins a spinner with 2 areas marked 50% and 2 areas marked 100%. Depending on what they spin, they either pay 50% or 100% of the price they bid. Play until everyone is out of money. Quarters or half dollars may need to be available to make change.

- **Show/Tell me what 50% of $2.00 is.**
- **Show/Tell me what 100% of $2.00 is.**
- **Show/Tell me what 50% of $1.00 is.**

Lesson Follow-Up

If students have difficulty completing this activity, have them complete Activity 39.

Activity 38

Independent **Objective**

Students round whole numbers to 500 to the nearest ten or hundred to determine a reasonable estimate in problem situations, and check for accuracy.

Talk to students about the fact that when you go shopping you need to know that you have enough money to pay for what you buy; however, you do not have to know exactly how much everything costs. This is when it is important to be able to estimate. Give each student a list of 3 items they might buy from a store and the price for each item. Tailor these items to the interests of the students in the class. The prices might be $2.75, $17.25, and $29.77. Explain to the students that when dealing with money, it is usually best to round up to the nearest dollar. This is to make sure there is always enough money. Work through each of the prices on the list with the class, rounding to the nearest dollar. Then add up the dollar amounts to estimate how much money they will need to buy the items.

- **Show/Tell me what $7.46 rounded to the nearest dollar is.**
- **Show/Tell me what $6.90 rounded to the nearest dollar is.**
- **Show/Tell me what $6.30 rounded to the nearest dollar is.**

Lesson Follow-Up

If students have difficulty completing this activity, have them complete Activity 40.

Conceptual Development Activities

Activity 39

Developing **Objective**

Students identify the value of money to $1.00 written as a decimal.

Make a set of cards with 10 pictures of coins adding up to a value of less than a dollar and 10 corresponding cards with the written amount of money. Turn the cards over. Pair the students up and have them play a matching game. When making the set of cards, take into account the level of the students. Making one set of cards that uses only dimes, one that uses only quarters, and one that uses only nickels will help students have more success.

- **Show/Tell me which card shows the value of 2 nickels.**
- **Show/Tell me which card shows the value of 3 nickels.**
- **Show/Tell me which card shows the value of 1 nickel and 1 dime.**

Lesson Follow-Up

If students have difficulty completing this activity, have them complete Activity 41.

Activity 40

Developing **Objective**

Students solve problems by counting and grouping to create sets of tens and ones to identify the value of whole numbers to 100.

NOTE: This activity might be easier as an individual activity.

Give each student 100 pennies. When it is his or her turn, have the student roll a number cube. Trade the student that many dimes for the appropriate number of pennies. For example, if a 4 is rolled give the student 4 dimes. The student gives you 40 pennies. The winner is the first student who trades in all of his or her pennies for dimes.

- **Show/Tell me how many dimes equal 40 pennies.**
- **Show/Tell me how many dimes equal 60 pennies.**
- **Show/Tell me how many dimes equal 10 pennies.**

Lesson Follow-Up

If students have difficulty completing this activity, have them complete Activity 42.

Activity 41

Emerging **Objective**

Students express and use quantities 1 to 7.

Have a bowl of counters on the table. Ask the students to take 2 counters. If the students take the correct amount of counters, congratulate them on being correct. Repeat this with various numbers up to 7.

Have several pencils on the table. Ask students the following:

- **Show/Give me 3 pencils.**

Repeat with other numbers up to 7.

Lesson Follow-Up

If students have difficulty completing the activity, focus on only one number. Then move on to a second number. Be sure to use a high preference item to help motivate the students.

Activity 42

Emerging **Objective**

Students solve problems by joining or separating sets of objects with quantities to 7.

Have a bowl of counters on the table. Ask the students to take out 2 counters. Then you take out 3 counters and put them on the table. Ask the students to combine the two piles of counters and tell you how many counters there are altogether. Help thc students push the two piles together if necessary. Initially, the students may need help counting the counters in the new pile. Repeat this with different numbers of counters in the two piles.

Have a pile of 3 pencils and a pile of 2 pencils.

- **Show/Tell me how many pencils there are altogether.**

Have a pile of 1 pencil and a pile of 2 pencils.

- **Show/Tell me how many pencils there are altogether.**

Have a pile of 2 pencils and a pile of 2 pencils.

- **Show/Tell me how many pencils there are altogether.**

Lesson Follow-Up

If students have difficulty completing the activity, focus on one to one correspondence and counting to 7. Then go back and work on joining groups together.

Conceptual Development Activities

Activity 43

Independent **Objective**

Students use data from a sample to make predictions regarding the whole group.

Tell the students they are going to figure out what the favorite colors are of students in the school. However, asking all the students in the school would take too much time, so you are going to show them a short cut. Choose 25 students from the school that will represent the school. Choosing a math class or two would be an easy way to do this. Have your students make a quick survey for these 25 students to fill out. Tally the answers. Set up a proportion to figure out how many students in the school like the colors based on the results of the 25 students surveyed.

- **Show/Tell me, if 50% of the students in your survey like blue, how many in the entire school like blue?**
- **Show/Tell me, if 25% of the students in your survey like white, how many in the entire school like white?**
- **Show/Tell me, if 70% of the students in your survey like white, how many in the entire school like white?**

Lesson Follow-Up

If students have difficulty completing this activity, have them complete Activity 45.

Activity 44

Independent **Objective**

Students match triangles that are similar (same shape but different size) using physical models.

Have the students look up the record for their three favorite sports teams, all within the same sport. The students can use the Internet, local papers, or an encyclopedia to find the information. Help them make a bar graph of the data. Each student can then create a 3-question quiz based on the data for you to take. The student must also supply an answer key. Students could also trade the quizzes among each other instead of giving them all to you.

Show students a bar graph of the number of games won by several professional football teams.

- **Show/Tell me, how many games did team _______ win?** Repeat question with a different team.
- **Show/Tell me, how many games were won by all the teams?**

Lesson Follow-Up

If students have difficulty completing this activity, have them complete Activity 46.

Conceptual Development Activities

Activity 45

Developing **Objective**

Students compare data shown in a pictograph with three categories and describe which categories have the largest, smallest, or the same amount.

Create a pictograph of how many wins the football, volleyball, and basketball team each have. Have students create a symbol for football, volleyball, and basketball. Have them make one symbol for each game won. Put them up on a wall or white board. Ask the students which team won the most games. Help students count the symbols if necessary. Then have them figure out which team won the fewest games. Have students brainstorm other teams or activities for which they could make pictographs.

Show the students a new pictograph.

- **Show/Tell me which team has won the most games.**
- **Show/Tell me which team has won the fewest games.**

Lesson Follow-Up

If students have difficulty completing this activity, have them complete Activity 47.

Activity 46

Developing **Objective**

Students use pictographs to display data in labeled categories and identify the number in each category.

Have students look up the record for their three favorite sports teams, all within the same sport. The students can use the Internet, local papers, or an encyclopedia to find the information. Help them make a pictograph of the data. Each student should label each pictograph and tell how many games were won by each team as represented by the pictograph.

Show students a pictograph of number of games won by several professional football teams.

- **Show/Tell me how many games team ______ won.** Repeat for different teams.
- **Show/Tell me how many games all the teams won.**

Lesson Follow-Up

If students have difficulty completing this activity, have them complete Activity 47.

Conceptual Development Activities

Activity 47

Emerging **Objective**

Students count the objects, pictures, or symbols use in a pictograph and identify total to 7 or more.

Create a pictograph showing the number of games the football team has won this year. Use the picture of a football to represent one win. Ask the students to count how many games the team has won. Repeat for pictographs of other local sports teams.

Have a pictograph of 5 football wins.

- **Show/Tell me how many games the football team won.**

Have a pictograph of 3 football wins.

- **Show/Tell me how many games the football team won.**

Have a pictograph of 4 football wins.

- **Show/Tell me how many games the football team won.**

Lesson Follow-Up

If students have difficulty completing the activity, focus on one- to-one correspondence and counting to 7. Then go back and work on counting the pictograph symbols.

Activity 48

Independent **Objective**

Students predict the likely outcome of a simple experiment and conduct the experiment to determine if prediction was correct.

Have a container with 12 marbles or colored counters, 6 blue, 2 red, 2 green, and 2 yellow.

- **If one marble is pulled out of a container, what color will it be?**

Have students write their predictions on the board. Pull out 1 marble and record the color.

- **If we put this marble back in the container and pulled out a marble, would the result be the same?**

Repeat the experiment multiple times. Remember to replace the marble that is pulled after each experiment. Discuss which color is most likely to be pulled out of the container.

- **Show/Tell me, if there are 12 marbles, 6 red, 2 blue, 2 green, and 2 yellow, what color is most likely to be pulled out of the container?**
- **Show/Tell me, if there are 12 marbles, 6 green, 2 blue, 2 red, and 2 yellow, what color is most likely to be pulled out of the container?**
- **Show/Tell me, if there are 10 marbles, 5 blue, 1 red, 2 green, and 2 yellow, what color is most likely to be pulled out of the container?**

Lesson Follow-Up

If students have difficulty completing this activity, have them complete Activity 49.

Conceptual Development Activities

Activity 49

Independent **Objective**

Students predict the likely outcome of a simple experiment by selecting from two choices and check to see if the prediction was correct.

Have students predict how many heads will turn up if a penny is flipped 10 times. Have the students write their predictions. Then let every student try the experiment. Have them compare their predictions to the actual result and then share their results. Have students repeat this experiment 10 times each. Again, have students share their results. Lead a discussion with the students about what would be the best guess for how many heads will show up. Try to get a large enough sample so 5 happens the most often.

- **Show/Tell me how many times heads will usually show up if a penny is flipped 10 times.**
- **Show/Tell me how many times tails will usually show up if a penny is flipped 10 times.**
- **Show/Tell me how many times heads will usually show up if a penny is flipped 6 times.**

Lesson Follow-Up

If students have difficulty completing this activity, have them complete Activity 50.

Activity 50

Independent **Objective**

Students recognize a common cause-effect relationship.

Make pictures that show some basic school rules. Examples might be a picture of a student running in the hall and a student walking in the hall, a student throwing food and a student eating lunch, or a student yelling in class and a student raising his or her hand. Show students one of the pictures and have them explain the outcome. Running in the hall would result in a negative consequence, although the exact answer will vary based on the students' behavior plan and school rules. Walking in the hall would result in a positive consequence. Discuss with the class the result of each behavior.

- **Show/Tell me what the result or effect is of running in the hallway.**
- **Show/Tell me what the result or effect is of raising your hand.**
- **Show/Tell me what the result is of throwing food.**

Lesson Follow-Up

If students have difficulty completing the activity, have them act out the scenarios. If video equipment is available, the scenarios can even be recorded and played back.

Conceptual Development Activities

Activity 51

Independent Objective

Students use information from physical models, diagrams, tables, and graphs to solve addition, subtraction, multiplication, and division equations.

Using a baseball box score from the newspaper, show students where to find the total number of hits and how many at bats each team had during the game. The abbreviation for the category at bats is AB; the abbreviation for hits is H. The total for each team appears after all players on the team have been listed. Ask the students to tell the total number of at bats throughout the game. They may use calculators. Students could also be asked to solve a variety of number sentences after mastering this task.

- **Show me where to find the total number of at bats Team A had in the game.**
- **Show me where to find the total number of at bats Team B had in the game.**
- **Tell me how many at bats there were during the entire game.**
- **Tell me how many more at bats Team A had than Team B.**
- **Tell me how many at bats Team A would have had if they doubled the number of at bats they had in this game.**
- **Tell me how many at bats Team B would have had if they halved the number of at bats they had in this game.**

Lesson Follow-Up

If students have difficulty completing this activity, have them complete Activity 54.

Activity 52

Independent Objective

Students identify the relationship between two sets of related data, such as ordered number pairs in a table.

Construct a table that reflects ordered pairs with a relationship. The first column could read 1, 2, 3, 4, 5, and the second column could read 6, 7, 8, 9, 10. Ask students to cover up all the pairs except the first. Ask them to show or tell how they would get from one number to the other. add 5 Reveal the second pair of numbers, and repeat the question. Repeat as necessary. Next, reveal only the first number of the next ordered pair. Ask students to determine what the second number of the ordered pair would be if they followed the pattern from the previous examples.

- **Tell me how you go from the first number to the second number in the ordered pair.**
- **Show/Tell the pattern these sets of numbers have.**
- **Tell me what number matches 3 in the first set.**

Lesson Follow-Up

If students have difficulty completing this activity, have them complete Activity 55.

Conceptual Development Activities

Activity 53

Independent **Objective**

Students translate problem situations into equations involving addition and subtraction of two-digit numbers and multiplication and division facts.

Create a bar graph with the high temperature recorded for each month of the year. Ask students to determine which month showed the highest temperature for that year. Ask the students to determine how much higher that month's temperature was compared to another. For example, if the month with the highest temperature was August, ask students how much hotter August's highest temperature was than March's. This will allow students to demonstrate the ability to translate information from the graph to a number sentence involving subtraction. A calculator is acceptable to determine the difference.

- **Show/Tell me the month that had the highest temperature.**
- **Show/Tell me the highest temperature for the month of March.**
- **Show/Tell me how much hotter it was in August than it was in March.**
- **Show/Tell me the month that had the lowest temperature.**

Lesson Follow-Up

If students have difficulty completing this activity, have them complete Activity 56.

Activity 54

Developing **Objective**

Students use information from physical models, diagrams, tables, and pictographs to solve equations involving addition and subtraction with one-digit and two-digit numbers.

Create a table showing three types of music students in the school prefer (rock, rap, or country). Conduct a survey if appropriate. Ask students to show or tell the number of students who liked each one of the styles. Ask students to tell how to determine the total number of students who voted. Ask students to tell how to determine the total number of students who voted for either rock or country.

- **Show/Tell me how many students voted for rock music.**
- **Show/Tell me how many more students voted for rock music than for country music.**
- **Show/Tell me how many students preferred rock and rap.**
- **Show/Tell me how many students voted in this survey.**

Additional activities to reinforece this activity could include the development of bar graphs for preferred fruit, preferred sports, preferred pizza toppings, and so on. Repeat this activity as needed.

Lesson Follow-Up

If students have difficulty completing this activity, have them complete Activities 57 and 58.

Conceptual Development Activities

Activity 55

Developing **Objective**

Students describe the relationship (1 more or 1 less) between two sets of related numbers.

Use a table to display sets of ordered pairs. The first column could read 8, 7, 6, 5, and 4; the second column would then read 7, 6, 5, 4, and 3. Cover all sets but the first pair (8, 7). Ask students to tell you if the second number in the pair is one more or one less than the first number in the pair. Reveal the second pair of numbers (7, 6), and repeat the question. Ask students if they see a pattern. Model the thought process aloud if students are unable to recognize the pattern. Reveal the first number in the third pair (6), and ask students what number will appear in the second column to correspond with the number 6.

- **Tell me if the number in the second column is more or less than the number in the first column.**
- **Show/Tell me the pattern that you see between column one and column two.**

Write the numbers 5, 6, and 7 for students to choose from.

- **Show me the number that will be in the second column if the number 6 is in the first column.**

Lesson Follow-Up

If students have difficulty completing this activity, have them complete Activities 57 and 58.

Activity 56

Developing **Objective**

Students translate real-world situations into equations involving addition and subtraction.

Make a pictograph showing how many days of rain there were in March, April, and May. Each symbol on the graph will represent 2 days of rain (2 clouds for March, 3 clouds for April, and 4 clouds for May). Ask students to show or tell how many rainy days were in March. Model the thinking process of reading the pictograph if necessary. Repeat the question for April and May. Model the process as necessary. If this activity is difficult, the pictograph could be used on a 1:1 basis (one picture equals one day of rain).

- **Show/Tell me how many rainy days there were if you see one cloud.**
- **Show/Tell me how many rainy days were in March.**
- **Show/Tell me how many rainy days were in April.**
- **Show/Tell me how many more rainy days were in May than in March.**
- **Show/Tell me the total number of rainy days in March, April, and May.**

Lesson Follow-Up

If students have difficulty completing this lesson, have them complete Activities 57, 58, or 59.

Conceptual Development Activities

Activity 57

Emerging Objective

Students solve simple real-world problems involving quantities using *more, less, smaller,* and *none.*

Using a deck of number cards, explain to students that each person needs five cards to play a game. Count out five cards with the students. Deal a second hand, this time stopping at four cards. Ask students if that is the same amount as the first stack of cards or if they will need more cards. Model the thought process for the students by lining up the cards from the first deal and matching them up with the second deal. When the students see one card without a match, tell them the second deal needs one more card. For the third deal, count out six cards to demonstrate that the hand has too many, and one less is needed.

- **Show/Tell me how many cards each person needs to play.**
- **Show/Tell if we need more cards in this hand.**
- **Show/Tell if we need more or less cards in this hand.**

Lesson Follow-Up

If students have difficulty with this activity, you can reduce the number of cards required for each hand. You could also have the students practice echoing you as they say "I have one card; I need more cards," until they reach the desired number of cards.

Activity 58

Emerging Objective

Students solve simple problems involving joining or separating sets of objects to 8.

Use photos or drawings of everyday items such as spoons and forks. Model for students how to separate the pictures into the two groups depending on which item they are. Next, hold the pictures for the students to see, and ask them to tell you which group the item belong to—spoons or forks. Repeat the exercise as necessary.

- **Show/Tell me which group this item belongs to.**

Place a fork in the spoons group.

- **Show/Tell me why this item does not belong in this group.**
- **Show/Tell me which item does not belong in this group.**

Lesson Follow-Up

If students have difficulty with this activity, hand-over-hand prompts may be used. Other sets of objects may also be used to ensure the student is familiar with the grouping.

Conceptual Development Activities

Activity 59

Emerging **Objective**

Students distinguish between the positions of two objects, such as *first* and *next*.

Outline two boxes on the floor using masking tape. Label the boxes 1st and 2nd. Play a game of musical boxes with the students. This is played just as musical chairs, except each person has a spot in which to go when the music stops playing. Go to the student in the first position and ask him or her to say "I am first." Ask this student to tell or show you who is next. If they are unable to answer, assist by saying the name of the student in the second position. Go to the student in the second position, and ask him or her to say "I am second." Ask this student to tell or show you who is first. Model the answer if necessary. Repeat the game with the same or with other students.

- **Tell me or show me who is first.**
- **Tell me or show me who is next.**

Adjust this activity to accommodate the needs of the students.

Lesson Follow-Up

If students have difficulty with this activity, you could have the other students in the class occupy the positions and take the student to each position physically.

Activity 60

Independent **Objective**

Students identify triangles that are similar (same shape but different size) using physical and visual models.

Give students pieces of paper that have been cut into squares. Each square should be a different size. Point out that even though they are different sizes, they are still the same shape. Ask students to fold the pieces of paper so that the bottom right corner meets the top left corner. Some students will have small triangles and some will have larger triangles. Ask students to line up their triangles either on the floor or on a wall from the smallest to the largest. Measure the sides of the first triangle, and record the measurements on the board. Measure and record the sides of the second triangle as well. Demonstrate the relationship between the bases of both triangles and the sides of both triangles. Continue with this exercise for as many triangles as necessary for the students to understand the correspondence. Introduce triangles with different shapes, and ask students to identify the triangles that have the same shape and those with a different shape.

- **Show/Tell me the relationship between the base of triangle A and the base of triangle B.**
- **Show me two triangles that have the same shape.**
- **Show me a triangle that has a different shape.**

Lesson Follow-Up

If students have difficulty with this activity, have them complete Activity 64.

Conceptual Development Activities

Activity 61

Independent **Objective**

Students form intersecting lines and identify the angles as *acute, obtuse,* or *right* by matching to a model.

Ask students to get out a piece of notebook paper. Ask a student to go up to the board and trace two connecting sides of their piece of paper, displaying a right angle. Draw lines from the vertex of that right angle to create other angles. Demonstrate that any angle that could be covered up by the paper would be called acute, which means the angle is less than 90 degrees. Ask another student to come up to the board and draw three lines that intersect. Point to one of the angles resulting from the intersections, and ask students to show or tell whether that angle is acute. They may use their pieces of paper if necessary. Repeat the exercise comparing obtuse and right angles.

- **Show/Tell me which of these angles are acute angles.**
- **Show/Tell me which of these angles are obtuse angles.**
- **Show/Tell me which of these angles are right angles.**

Lesson Follow-Up

If students have difficulty completing this activity, have them complete Activity 65.

Activity 62

Independent **Objective**

Students distinguish angles within triangles as *acute, obtuse,* or *right* using a right angle as a model.

Ask students to get out a piece of notebook paper. Ask a student to go up to the board and trace two connecting sides of their piece of paper, displaying a right angle. Review with students that a right angle is 90 degrees. Use a straightedge to connect the lines the student drew, creating a triangle. Next to the figure, write 180. Explain that a triangle can be composed of only 180 degrees of angles. Write 90 under the 180, and include a subtraction sign. If a triangle has 180 degrees and one of the angles is 90 degrees, that only leaves 90 degrees for both the two other angles. Have students test the other angles by holding their piece of paper up to them. If one of the lines is less than the right angle of the paper, it is acute. A protractor could also be introduced for this lesson, if appropriate.

- **Show/Tell me the sum of the angles of a triangle.**
- **Show/Tell me if this angle is acute, right, or obtuse.** Repeat for all three angles in the triangle.

Lesson Follow-Up

If students have difficulty completing this activity, have them complete Activity 66.

Conceptual Development Activities

Activity 63

Independent **Objective**

Students locate the right angle and the hypotenuse in a right triangle.

Ask students to make a list of the right angles they can find in the room. There can be prizes for students if they have at least 10 different examples. After they have completed their lists, find the most convenient of these examples to create a triangle. Re-create the right triangle on the board to make it easier to record the measurements. Ask students to measure the sides of the triangle, and record them on the board. Explain that the hypotenuse is always the side opposite the right angle and is always the longest side. Have students prove this with three of the examples from their list of right angles. They will need to re-create their triangles on paper just as you did together on the board.

- **Show/Tell me what the longest side of a right triangle is called.**
  a. hypotenuse b. hyperbole c. perimeter
- **Show me the hypotenuse.**
- **Show me the right angle.**

Lesson Follow-Up

If students have difficulty completing this activity, have them complete Activity 67.

Activity 64

Developing **Objective**

Students match triangles that are similar (same shape but different size) using physical models.

Give students pieces of paper that have been cut into triangles. Each triangle should be a different size. Point out that even though they are different sizes, they are still the same shape. Ask students to display their triangles either on the floor or on a wall. Prepare triangles of other shapes ahead of time, and add them to the display the students have made. Take the first triangle, hold it in front of the next in line, and ask the students to tell you if the two are the same shape. Model the thought process if students are struggling. Repeat the exercise with the rest of the triangles in the line.

- **Show/Tell me which two triangles have the same shape.**
- **Show/Tell me two triangles that are different.**

Lesson Follow-Up

If students have difficulty completing this activity, have them complete Activity 68.

Conceptual Development Activities

Activity 65

Developing **Objective**

Students identify angles formed by intersecting lines.

Draw angles of varying sizes on paper with students assisting. Have students cut out the angles. They will be used as models of what an angle looks like. On the board, draw a pair of intersecting lines. Ask students to see whether they can match any of the angles they cut out to the lines on the board. If they have difficulty, assist them in matching the proper angle. When all the angles have been found, repeat the exercise. On the third round, ask students not to use their cut-outs. Ask students to come up to the board and point to one angle. When an angle is identified, use a different color inside the angle to show that it is no longer an option for another student. Repeat as necessary.

- **Show/Tell me what an angle looks like.**
- **Show/Tell me where I can find an angle in these lines.**
- **Show/Tell me another angle.**

Lesson Follow-Up

If students have difficulty completing this activity, have them complete Activity 69.

Activity 66

Developing **Objective**

Students identify the angles within a triangle.

Have several lines prepared (laminated with either end beveled) that can be attached by magnets or tape to a whiteboard or other surface. If necessary, stick the lines to the board. Ask a student to use two of the lines to create an angle. Take a third line, and add it to the figure on the board to create a triangle. Have the student show or tell you the angles within the triangle. If this is difficult for the student, move one of the lines slightly away from the figure to show one angle, and mark it so the student can see it easily. Move the line back to restore the triangle. Repeat the process to show the other two angles. Repeat the exercise using differently sized and differently oriented triangles.

- **Show/Tell me how many angles a triangle has.**
- **Show/Tell me how many lines make up an angle.**
- **Show/Tell me how many lines make up a triangle.**

Lesson Follow-Up

If students have difficulty completing this activity, have them complete Activity 69.

Conceptual Development Activities

Activity 67

Developing **Objective**

Students locate the right angle within a right triangle.

Ask students to get out a piece of notebook paper. Ask a student to go up to the board and trace two connecting sides of their piece of paper, displaying a right angle. Label the angle. Take students on a tour of the room, comparing the right angles found all around to the one drawn on the board. Examples can include tables or table legs, corners, walls or floor, and so on. Bring students' attention back to the board. Use a straightedge to connect the two lines drawn on the board, creating a triangle. Ask students to guess the name of that triangle. If they cannot name the triangle, cover the third line. Remind them that the angle they drew is called a right angle. Uncover the third line, and model the thought process of coming up with the name *right triangle*. Draw other right triangles, being sure to label the right angle, and ask students to identify the right angle. Repeat as necessary.

- **Show/Tell me where the right angle is in this triangle.**

Lesson Follow-Up

If students have difficulty completing this activity, have them complete Activity 70.

Activity 68

Emerging **Objective**

Students recognize a triangle.

Have a triangle available to demonstrate to a student. Tell students the properties of a triangle. **Triangles have three straight sides that connect to make three corners.** Trace the triangle with your finger, and count the sides with the student. Have the student trace the triangle as you both count the sides. Repeat the exercise with a triangle of a different size or shape. Place a triangle and a square on the table, and ask the student to point to the triangle. If he or she is not able to do this, have the student trace the square with their finger, and both of you count the sides. Review that triangles have three sides. Again, ask the student to point to the triangle. As the student progresses, add more figures to the table.

- **Show/Tell me how many sides a triangle has.**
- **Show me or tell me which one is the triangle.**

Lesson Follow-Up

If students have difficulty with this activity, do not vary the size or shape of the triangle until they are able to identify it with consistency against one other shape. Using a marker to trace the triangle would demonstrate where the student should start or stop counting.

Conceptual Development Activities

Activity 69

Emerging **Objective**

Students recognize corners and angles in two-dimensional shapes, including rectangles and triangles.

Have several lines prepared (laminated with either end beveled) that can be attached by magnets or tape to a whiteboard or other surface. If necessary, stick the lines to the board. Ask a student to use two of the lines to create an angle. If this is difficult for a student, assist them. When the angle is on the board, have the student trace it with their finger as you both count the sides and note the corner. Take a third line, and add it to the figure on the board to create a triangle. Have the student trace the figure again, counting the sides and all three corners. Ask the student to show you all the corners in the triangle. He or she could use a marker to mark the angle. Ask the student to show you all the angles in the triangle.

- **Show/Tell me how many corners are in an angle.**
- **Show/Tell me how many angles a triangle has.**

Lesson Follow-Up

If students have difficulty answering these questions, the laminated lines can be separated slightly from the figure to show the different angles and corners as necessary.

Activity 70

Emerging **Objective**

Students recognize the hypotenuse as the longest side of a triangle.

Ask students to get out a piece of notebook paper. Ask a student to go up to the board and trace two connecting sides of their piece of paper, displaying a right angle. Label the angle. Use a straightedge to connect the lines the student drew, creating a triangle. Ask students to guess which of the lines is the longest. Ask students to use a yardstick or ruler to measure the sides of the triangle to see if anyone guessed correctly. Make sure to record the lengths of all the sides. Demonstrate for the students that the longest side is opposite the right angle. Repeat the exercise using larger or smaller pieces of paper to begin the triangles. When the students consistently begin to recognize the hypotenuse as the longest side, vary the orientation of the triangles, but continue to label the right angle.

- **Show/Tell me where the right angle is.**
- **Show/Tell me which side is the longest.**
- **Show/Tell me what the longest side of a right triangle is called.**
  a. perimeter b. area c. hypotenuse

Lesson Follow-Up

If students have difficulty, continue to measure the sides of the triangles, making sure to record the results. The hypotenuse and the right angle could be color-coded to display a relationship between the two.

Conceptual Development Activities

Activity 71

Independent Objective

Students organize data and display it in bar and line graphs.

Tell students they are to become coaches of a professional sports team, and their first job is to look at the win-loss record for the team for the past several years. Provide them with the statistics for about five years; for example, in 2008 the team was 11–5. Tell them that to view all the information easily, they will be making a graph. On a whiteboard, draw the vertical and horizontal axes. Ask students for suggestions on the title of the graph. Record the agreed-upon title. To the side of the axes, list the information to be shown on the graph, such as year and number of wins. Model the thinking process as you number the vertical axis and horizontal axis. Ask students to continue with the rest of the years available. When the graph is completed, ask students to create a graph for the losses, or add the losses to the first graph.

Ask students the following:

- **Using this graph, show/tell me which year the team won ______ games.**
- **Using this graph, show/tell me which year the team won the most games.**
- **Using this graph, show/tell me how many more games the team won in 2005 than in 2006.**

Lesson Follow-Up

If students have difficulty completing this activity, have them complete Activity 73.

Activity 72

Independent Objective

Students determine the largest number, the smallest number, the mode, and the median of a set of data with up to 9 numbers.

Give a number cube to a student, and ask the student to record on the board what they rolled. Roll the number cube five times. Have students help you arrange the numbers on the board from the smallest to the largest. Be sure to cross out the numbers in the first set as they are recorded in the second set.

- **What was the smallest number?**
- **What was the largest number?**
- **What number came up most often?**

Explain to the students that the number that appears the most often is called the mode. Put the number in order from smallest to largest.

- **What number is in the middle?**

Students can work individually or in pairs to roll dice seven times to create a new set of numbers. Ask students to reorder the set they rolled and to identify largest and smallest number and to identify the mode and median of their sets.

- **Show/Tell me the smallest number rolled.**
- **Show/Tell me the largest number rolled.**
- **Show/Tell me the number rolled most often.**
- **Show/Tell me the middle number.**

Lesson Follow-Up

If students have difficulty completing this activity, have them complete Activity 74.

Conceptual Development Activities

Activity 73

Developing **Objective**

Students organize data in pictographs.

Have magnets or laminated shapes prepared that can be attached to a poster or a whiteboard. Shapes should be further distinguished by color-coding. Students could take a short survey to find out a class's favorite candy, or you could provide the results of a survey. Ask students to help organize the information about the candy. Draw the horizontal and vertical axes on a whiteboard or poster the students will be able to reach. List the information to be depicted on the graph. Ask the following questions: **What should we title the graph? Where should the numbers go? What should represent the suckers? The chocolate bars? The gum?** After these questions are answered, ask students to attach the symbols to the graph in the proper places. When students can consistently work with the pictographs as a 1:1 ratio, a 2:1 ratio could be introduced.

- **Show/Tell me how many students chose chocolate as their favorite candy.**
- **Show/Tell me how many more students like chocolate than the gum.**
- **Show/Tell me how many students like either chocolate or gum.**

Lesson Follow-Up

If students have difficulty completing this activity, have them complete Activity 75.

Activity 74

Developing **Objective**

Students identify the mode in a set of data with up to 5 numbers.

Give a number cube to a student, and ask the student to record on the board what he or she rolled. Roll the number cube five times. Have students help you arrange the numbers on the board from the smallest to the largest. Be sure to cross out the numbers in the first set as they are recorded in the second set.

- **What number came up most often?**

Circle the mode. Repeat the exercise.

- **Show/Tell me how many times the number 2 was rolled.**
- **Show/Tell me which number was rolled the most.**

Lesson Follow-Up

If students have trouble completing this activity, have them complete Activity 75.

Conceptual Development Activities

Activity 75

Emerging **Objective**

Students count the objects, pictures, or symbols used in a pictograph or chart and identify a total to 8.

Have a few pictographs prepared either on a whiteboard or handouts, using items or subjects with which the students are familiar. For example, use a pictograph showing the favorite candy of the class: chocolate, suckers, and gum. Use symbols that could be easily associated with each kind of candy, such as brown for chocolate, and so on. Ask students to count with you to see how many chose gum as their favorite candy. Touch each symbol as the counting progresses. Repeat with the next candy. Ask the students to count the final candy on their own. Introduce the next pictograph. Ask the students to count with you for the first category only, continuing by themselves.

- **Show/Tell me how many students chose chocolate as their favorite candy.**
- **Show/Tell me how many students chose gum as their favorite candy.**
- **Show/Tell me how many students chose either gum or suckers as their favorite candy.**

Lesson Follow-Up

If students have difficulty completing this lesson, continue to count along with the students. Hand-over-hand prompts may also be necessary for some students.

Activity 76

Independent **Objective**

Students identify the meaning of the variables in stated formulas involving area and perimeter.

Hand out graph paper. Tell students that they are to create blueprints for their dream school. They must include a few features of a normal school but may add features they want to include, such as a skate park or swimming pool. They are to draw and label rectangular rooms using the line of the graph paper. When they are finished, tell students they need to find the area of the rooms in order to buy flooring. Use a room one of the students has created to find the area as a group. Write the formula for area on the board: *Area* = *length* × *width*. Draw a representation of the room on the board with the letters l and w as substitutes for *length* and *width*. Ask students to count the boxes horizontally and vertically along the room; write those numbers on the board as well. Model the thought process aloud as the group works through the problem. When they are finished with the first room, ask students to find the area for three more of their rooms.

- **Show/Tell me what l stands for in the equation $A = l \times w$.**
- **Show/Tell me what w stands for in the equation $A = l \times w$.**
- **Show/Tell me what A stands for in the equation $A = l \times w$.**

Lesson Follow-Up

If students have difficulty completing this exercise, have them complete Activity 78.

Conceptual Development Activities

Activity 77

Independent **Objective**

Students translate real-world problem situations into equations and inequalities involving addition, subtraction, and multiplication.

Create a table and a bar graph that show how many new tricks a skateboarder has learned each month. Prepare several questions about the graph. Demonstrate the first question for the students.

- **How many more tricks did Jimmy learn in May than he did in April?**

Model the thinking process aloud. Say, **I know that when the problem asks *how many more,* I am going to use subtraction. If Jimmy learned 8 tricks in May** (write 8 on the board), **and he learned 5 tricks in April** (write that on the board), **I have to subtract the 5 from the 8** (add the subtraction sign). Ask students to solve the equation. Echo their answer, and include the phrasing from the question. Say, **Jimmy learned 3 more tricks in May than he did in April.** Ask students to complete the other questions.

- **Show/Tell me the correct equation to solve how many tricks he learned in April and May.**
  a. $5 - 8 =$ _____
  b. $5 + 8 =$ _____
  c. $8 - 5 =$ _____
- **Show/Tell me how many tricks he learned all year.**
- **Jimmy's friend Ariel learned 3 times as many new tricks in the month of April as Jimmy did. How many new tricks did Ariel learn?**

Lesson Follow-Up

If students have difficulty completing this activity, have them complete Activity 79.

Activity 78

Developing **Objective**

Students demonstrate how to determine the perimeter of figures such as rectangles.

Tell students it is their job to create a yard to that will be used to keep animals for a petting zoo. Use wooden counting sticks (with units marked) to construct a rectangle on the floor in a space students are able to see. Tell students that we have to know how much wood to purchase to construct the fence. Have students count the units on the wooden counting sticks for each side of the rectangle. Record on the board the measurement for each side. Help students add the numbers to reach the end result. Repeat the exercise, this time having the students create the rectangle. If appropriate, have students use a yardstick or tape measure to measure the fence.

- **Here we have a model of a rectangle. Show/Tell me how many units we need to construct this side of the rectangle.**
- **Show/Tell how many units we need for the entire rectangle.**

Lesson Follow-Up

If this activity is difficult for students to complete, have them complete Activities 80 and 81.

Conceptual Development Activities

Activity 79

Developing **Objective**

Students translate real-world problems situations into equations involving addition and subtraction of one- and two-digit numbers.

Create a pictograph with detachable symbols that shows how many new tricks a skateboarder has learned each month.

- **How many more tricks did Jimmy learn in May than in June?**

Have students find both of those months on the pictograph. Beginning with May, have students count or pull off the symbols for each month and record them on the board. Model the thinking process aloud. Say, **I know that when the problem asks *how many more*, I am going to use subtraction.** Add the subtraction sign. Use the symbols from the graph to demonstrate solving the equation. Repeat the process for two other months.

- **Show/Tell me which operation to use in this problem.**
  a. multiplication
  b. subtraction
  c. addition
- **Show/Tell me the correct equation to solve this problem.**
  a. $5 - 8 =$ ____
  b. $5 + 8 =$ ____
  c. $8 - 5 =$ ____
- **Show/Tell me the answer to the equation $8 - 5 =$ ____.**

Lesson Follow-Up

If students have difficulty completing this activity, have them complete Activities 80 and 81.

Activity 80

Emerging **Objective**

Students identify a given quantity to 7 and add 1 more to solve problems.

Have a number line posted for the students to see and reach, if appropriate. Use simple problems that name items familiar to the students. For example, say: **If we have three chairs and we need one more, how many do we need?** Write the equation on the board. Count out three on the number line with the students, and count three actual chairs in the room. Demonstrate moving an additional space on the number line when adding one more, and add the actual object to the area. Ask students to name the number where the counting stopped. Repeat the exercise with other numbers.

- **Show me or tell me how many chairs we have to start.**
- **Show me or tell me how many more chairs we need.**
- **Show me or tell me how many chairs we will have at the end.**

Lesson Follow-Up

If students have difficulty completing this lesson, use hand-over-hand prompts while counting. Do not vary the objects until the concept is demonstrated consistently.

Conceptual Development Activities

Activity 81

Emerging **Objective**

Students identify a given quantity to 8 and take away 1 to solve problems.

Use simple problems that name items familiar to the students. For example, colored counters could be used to demonstrate this operation. Set out up to 8 counters, and place them on the table. Have students count them out aloud as they point to each one. Allow a student to put one counter back in the container. Ask students to count how many are left after one is taken away. Count aloud with the students, and repeat the equation when finished. Say, **8 minus 1 equals 7.** Repeat the exercise, allowing the next student to remove one of the counters. Each time a counter is put back in the container, state the equation students are solving.

Ask students the following:

- **Show/Tell me how many counters we have.**
- **Show/Tell me how many counters we have left after we take away 1.**

Lesson Follow-Up

If students have difficulty completing this activity, assist them with hand-over-hand prompts. Matching the counters onto a number line may be useful.

Activity 82

Independent **Objective**

Students convert measures within the same system, including money, length, time, and capacity.

Have students keep a log of which television shows they watch or what games they play each day for a week. Have them record the length of the show, or the time spent with the games in minutes. Have a log prepared to use as an example for the class. Ask students to help convert the time into hours. They may use calculators. Ask them to add the total number of minutes spent on those activities. Ask the students to suggest how to change minutes to hours. If this is difficult, draw circles on the board to represent clocks. Each time a circle is filled, subtract 60 from the total number of minutes. When all the circles are full, show the students how they could use division to solve the problem. Ask the students to use the information from their logs to convert the minutes into hours.

- **Show/Tell me how many minutes are in one hour.**
- **Show/Tell me how many hours you have if you have 120 minutes.**

Lesson Follow-Up

If students have difficulty completing this activity, have them complete Activity 83.

Conceptual Development Activities

Activity 83

Developing **Objective**

Students identify standard units of measurement for length, weight, capacity, and time.

Construct the figure of "Gallon man" as a class. Prepare the different parts of Gallon Man to be pieced together by the class. Make sure to color code the different measures. They could reflect the colors of a favorite team in the area, such as making the cups blue, the pints orange, and so on. Have the cups as fingers, the pints as wrists, the quarts as arms, leading to the center of the body or the gallon. Repeat this construction for another arm and two legs. Have the students count the appendages of "Gallon Man" to identify the units involved in making a gallon. A simple recipe, of lemonade, perhaps, could be used to reinforce the different units using the figure.

- **Show/Tell me which unit is larger.**
  a. a pint **b. a gallon**
- **Show/Tell me how many quarts make a gallon.**
- **Show/Tell me how many cups make a quart.**

Lesson Follow-Up

If students have difficulty completing this activity, have them complete Activity 84.

Activity 84

Emerging **Objective**

Students recognize tools used for measurement, such as clocks, calendars, and rulers.

Collect a clock and a calendar. Place them where all students can see and reach. Prepare several flashcards with things associated with either of these two categories such as a vacation, a TV show, and so on. Hold up the flashcards, and read aloud with the students. Ask a student to place the flashcard under the item that would be used to measure the words on the flashcard. For example, *a vacation* would go under the calendar and *a TV show* would go under the clock. If students have difficulty, eliminate one measurement apparatus and continue with the flashcards, dealing only with the other instrument. Repeat as necessary.

- **Show/Tell me something that a clock measures.**
- **Show/Tell me something a calendar measures.**
- **Show/Tell me which would be the best item to measure a TV show.**

Lesson Follow-Up

If students have difficulty with this activity, review the units of time before proceeding with the exercise.

Conceptual Development Activities

Activity 85

Independent **Objective**

Students express, represent, and use whole numbers to 1,000 in various contexts.

Tell the class that they all have suddenly become millionaires. They do not have to worry about money, but they have to learn to keep track of what they spend to keep the accountants happy. They will also have to learn to write checks. Have several mock checks available for them to use. They will each receive a ledger to record their spending, and a list of items they will be able to purchase with the cost of those items. All items should be $1000 or less. Demonstrate for the students how to write a check, with the amount as both a number and the number in words on the check. Ask students to choose three items they would like to purchase. Require students to write the checks and record them on the ledger. After they complete their purchases, ask students to total their spending. They may use a calculator.

Students could also look up items on the Internet or in a newspaper advertisement.

- **Show/Tell me how to write 900 in words.**

Lesson Follow-Up

If students have difficulty completing this activity, have them complete Activity 89.

Activity 86

Independent **Objective**

Students round whole numbers to 1000 to the nearest ten or hundred to determine a reasonable estimate in problem situations, and check for accuracy.

Talk to the students about being able to obtain their driver's licenses in a couple of years. Ask them to think about all the things they will need to have for a car, such as insurance, gas money, maintenance, and so on. Allow the students to pick one of those needs and research the monthly cost. Explain that they are to estimate the cost over two months. In order to estimate, demonstrate rounding to the nearest tens or hundreds. Ask the students to estimate how much insurance would be for two months. Have students explore the actual cost and the estimated cost. Ask students to research more of the expenses and estimate the total monthly cost of owning the car.

- **Here is the number 146. Show/Tell me what number I would have if I rounded to the nearest ten.**
- **Show/Tell me an estimated cost of insurance for 2 months.**
- **Show/Tell me the actual cost of insurance for 2 months.**

Lesson Follow-Up

If students have difficulty completing this activity, have them complete Activity 90.

Conceptual Development Activities

Activity 87

Independent **Objective**

Students express, represent, and use fractions, including halves, fourths, thirds, eighths, and sixths.

Ask the students to help you solve a problem. You need to have equal shares of pumpkin seeds for each student in Science class. Have enough seeds for the students to end with an equal amount. Count out the seeds, and write that number on the board. Below that number, write the fraction $\frac{1}{2}$. Ask one student to join you, and tell the class that one half of the seeds need to go to that student. Demonstrate using division to solve the problem, such as dividing the number of seeds by 2. Write the answer on the board. Ask students to check your work by physically moving the seeds into 2 piles—one in front of the student, one in front of you. Ask another student to join you and the first student. Continue the activity with halves. Repeat for up to eight students.

- **Here we have 32 seeds. Show/Tell me how many seeds equal one half of the total seeds.**
  a. 15 **b. 16** **c. 17**
- **Show/Tell me how to write *one half*.**
  a. $1\frac{1}{2}$ **b. $\frac{11}{2}$** **c. $\frac{1}{2}$**
- **Show/Tell me how many piles we would need if we wanted to find one third of the total seeds.**
  a. 1 **b. 2** **c. 3**

Lesson Follow-Up

If students have difficulty completing this activity, have them complete Activity 91.

Activity 88

Independent **Objective**

Students express, represent, and use percents-including 25%, 50%, 75%, and 100%- and decimals in the context of money.

Before doing this activity, cut out a colored circle to represent a decimal point. Write 100 on the board. Next to it, write 100%. Show students where to place the decimal point in the proper position to show 100% of a dollar. Place the decimal point between the 1 and the first 0. Leave this example on the board. For the second example, write 90 on the board, and next to it write 90%. Again, demonstrate for students where to place the decimal point in the proper position to show 90% of a dollar. Leave this example on the board as well. Write 75 and 75% on the board. Have students demonstrate where to place the decimal using the colored circle. Continue this activity using 50% and 25%.

Demonstrate how to use a calculator to determine percent if appropriate.

- **Show/Tell me how much 25% of a dollar is.**
  a. $1.00 **b. $0.50** **c. $0.25**
- **Show/Tell me what percentage of a dollar $0.50 is.**
  a. 100% **b. 50%** **c. 25%**
- **Show/Tell me what 100% of $100 looks like.**
  a. $10.00 **b. $1.00** **b. $100.00**

Lesson Follow-Up

If students have difficulty completing this activity, have them complete Activity 92.

Conceptual Development Activities

Activity 89

Developing **Objective**

Students express, represent, and use whole numbers to 100 in various contexts.

Ask students to pretend they are going shopping. Prepare pictures of the items for students to purchase. Each of the items will have its cost written in word form under the picture. The students will each be given play money up to one hundred dollars. To begin the exercise, bill denominations should match the cost of some items. Ask students to purchase three items with their bills. Ask students to count out the bills for their purchases. If students master this procedure, the cost of the items could be varied so that students must use combinations of bills to purchase some items.

- **Show/Tell me what you can buy with a $5 bill.**
- **Show/Tell me how much the radio costs.**
- **The game costs thirty dollars. Show/Tell me which bills you can use.**

Lesson Follow-Up

If students have difficulty completing this lesson, have them complete Activity 93.

Activity 90

Developing **Objective**

Students use counting, grouping, and place value to identify the value of whole numbers to 100.

Divide a table into three areas. Below the table, put jars containing play money that represents pennies, dimes, and dollar bills—one jar under each area. Ask students to place a dime on the table in the center area. Next, ask students to gather enough pennies on the table that equal the value of the dime. Finally, ask a student to circle with their finger the group of pennies that equals ten cents. Clear the table, and try the exercise again, this time using 2 dimes. When students are consistently successful, begin to use the dollar bills with the dimes. Using all three denominations would be a further challenge.

Ask students the following:

- **Here we have one dime. Show/Tell me how many pennies equal one dime.**
- **Show/Tell me how many things are in the set of dimes that equals ten cents.**
- **Show/Tell me how many things are in the set of pennies that equals ten cents.**

Lesson Follow-Up

If students have difficulty completing this activity, have them complete Activity 94.

Conceptual Development Activities

Activity 91

Developing Objective

Students express, represent, and use fractions, such as halves, fourths, and thirds.

Start with a picture of the top view of a birthday cake. Make four copies of the picture, and divide up three of the pictures, one in half, one in thirds, and one in fourths. The pieces of the divided pictures could be attached to poster board with Velcro. Write the respective fractions ($\frac{1}{1}$, $\frac{1}{2}$, $\frac{1}{3}$, and $\frac{1}{4}$) on the board next to the appropriate pictures. With the first picture, ask **Is the cake whole, or is it divided into pieces?** When the students reach the correct answer, point to the fraction $\frac{1}{1}$ to highlight the denominator. Move on to the second picture. Ask students if this cake is whole or divided into pieces. Ask how many pieces the cake is divided into. Circle the denominator of the fraction $\frac{1}{2}$ with your finger to show the number 2. Circle the two pieces of the cake individually. Ask a student to come to the board and show what it would look like if only half the cake were there. Assist the student if necessary. Continue with the other pictures.

- **Show/Tell me what the fraction one-half looks like.**
- **Here is a picture with two equal pieces. Show/Tell me what one-half of the picture looks like.**
- **Show/Tell me what the bottom number is in the fraction *one-third*.**

Lesson Follow-Up

If students have difficulty completing this activity, have them complete Activity 95.

Activity 92

Developing Objective

Students identify percents including 50% and 100%.

Start with a picture of the top view of a birthday cake. Divide the picture into 2 equal parts. Attach the two pieces on a poster board with Velcro. Circle the entire cake with your finger, and ask students to tell you how much of the cake is pictured. Explain that all of the cake is pictured, so 100% of the cake is shown. Write the number 100 next to the cake. Use a colored circle as the decimal point after the 1. Show students to move the decimal point two spaces to the right to show 100%. Ask students what the cake would look like if we had only half of the cake. Take away one portion of the cake. If we have $\frac{1}{2}$ of the cake, we are dividing the number one by the number two to get a decimal. Have or help students to use a calculator to get the answer. Write the answer 0.50 on the board. Ask a student to show you how to make 0.50 into a percentage. Have or help the student move the decimal point two spaces to the right. Have the students identify other objects as being 100% or 50%.

- **Here we have a cake with 2 equal halves. Show/Tell me how many halves we need to have 100% of the cake.**
- **Here we have a cake with 2 equal halves. Show/Tell me how many halves we need to have 50% of the cake.**

Lesson Follow-Up

If students have difficulty completing this activity, have them complete Activity 95.

Conceptual Development Activities

Activity 93

Emerging **Objective**

Students identify quantity in sets to 8.

Ask students to help you organize the change in a jar to see if you have enough money to buy lunch for the week. Have a jar with play quarters prepared for the activity. Tell the students that lunch costs $2 each day.

- **How many quarters do we need for lunch?**

Assist the students in processing by saying **I know four quarters equals $1. How many do I need for $2?** Help students count out the first four quarters, followed by the second. Each time a dollar is made, circle the group with your finger to show it as a set.

- **Show me or tell me how many quarters I need to make $1.**
- **Show me or tell me how many quarters I need to make $2.**
- **Show me or tell me how many quarters I need to buy lunch on Tuesday.**

Lesson Follow-Up

If students have difficulty with this activity, reduce the amount necessary to purchase lunch. Have the students circle each set they make and repeat value of the set. Hand-over-hand prompts may be useful.

Activity 94

Emerging **Objective**

Students demonstrate one-to-one correspondence by counting objects or actions to 8.

Have or help students assemble in a circle. Explain that they are going to play a game. For each round of the game, each student will have one turn. The card game "War" could work for this exercise. Lead the activity by putting one card in the center of the group. Verbalize the action. Say, **I am one player, and I put down one card.** Ask the next student to take their turn. Verbally reinforce the action they take when done correctly. When all students have taken their turn, count the players with the students, and count the number of cards that have been played. If students have been successful in following directions and the numbers match, determine the winner of that round. If the numbers do not match, try the round again, assisting students when necessary.

- **Here we have one player. Show me or tell me how many cards the player should put down.**
- **Here we have four players. Show me or tell me how many cards we should see at the end of the round.**
- **Here we have the second player. Show me or tell me how many cards this player should put down.**

Lesson Follow-Up

If students have difficulty with this activity, assist them with hand-over-hand prompts as their turn arrives. This activity could also be done in partners to reduce the number of players.

Activity 95

Emerging **Objective**

Students recognize half and whole sets of objects to 8.

Start with a picture of the top view of a birthday cake. First, divide the picture in half. Next, divide the picture into 8 equal parts. Assemble the pieces on a poster board with Velcro. Circle the entire cake with your finger, and ask the students to tell you how many pieces are in the picture. Now, ask students to count the pieces of the cake with you as you take them off the board. Tell students that you want to eat half of the cake today and save half the cake for tomorrow. Ask students to help you put the correct amount of pieces back on the board to show the half to be eaten tomorrow. Help the students divide the pieces, half on the board and half in a pile on the side. When the pieces are divided, help the students count each set to show that the piles are equal. Repeat the activity with the four pieces left on the board, allowing students to act more independently in this round.

- **Show/Tell me how many cakes are shown in this picture.**
- **Show/Tell me how many pieces of cake are in the picture.**
- **Show/Tell me how many pieces of cake equal half of the cake.**

Lesson Follow-Up

If students have difficulty completing this activity, begin with the cake divided into fewer portions, and use hand-over-hand prompts when counting out the equal piles.

NOTES